Plants of Central Asia
Plant Collections from China and Mongolia

Volume 6

T0199553

Plants of Central Asia

Plant Collections from China and Mongolia

Volume 6
Equisetaceae—Butomaceae
Bibliography (Supplement 1)

V.I. Grubov
A.E. Matzenko
M.G. Pachomova

CRC Press
Taylor & Francis Group
Boca Raton London New York

CRC Press is an imprint of the
Taylor & Francis Group, an **informa** business

A SCIENCE PUBLISHERS BOOK

ACADEMIA SCIENTIARUM URSS
INSTITUTUM BOTANICUM nomine V.L. KOMAROVII
PLANTAE ASIAE CENTRALIS
(secus materies Instituti botanici nomine V.L. Komarovii)
Fasciculus 6
EQUISETACEAE-BUTOMACEAE, BIBLIOGRAPHIA (ADDENDA 1)
V.I. Grubov, A.E. Matzenko and M.G. Pachomova conficerunt

First published 2002 by Science Publishers Inc.

Published 2019 by CRC Press
Taylor & Francis Group
6000 Broken Sound Parkway NW, Suite 300
Boca Raton, FL 33487-2742

© 2002, Copyright reserved
CRC Press is an imprint of Taylor & Francis Group, an Informa business

First issued in paperback 2019

No claim to original U.S. Government works

ISBN 13: 978-0-367-44711-3 (pbk)
ISBN 13: 978-1-57808-117-2 (hbk)
ISBN 13: 978-1-57808-062-5 (Set)

Visit the Taylor & Francis Web site at
http://www.taylorandfrancis.com

and the CRC Press Web site at
http://www.crcpress.com

    Library of Congress Catalogoing-in-Publication
Rasteniia TSentral'noi Azii. English
  Plants of Central Asia: plant collections from China and
  Mongolia
  /[editor-in-chief V.I. Grubov].
      p.   cm
  Research based on the collections of the V.L. Komarov
  Botanical Institute.
  Includes bibliographical references.
  Contents: V. 6 Equisetaceae-Butomaceae, Bibliography
  Supplement 1)
  ISBN 1-57808-117-3   (v.6)
        1. Botany-Asia, Central. I. Grubov V. I. II.
  Botanicheskiî institut im. V.L. Komarova. III. Title.
QK374, R23613 1999
581.958-dc21   99-36729
                                            CIP

Translation of: Rasteniya Central'noy Asii, Vol. 6, 1963
               Nauka Publishers, Leningrad.

NOTES

UDC 582.374/.538 (51)

PLANTS OF CENTRAL ASIA. From the Material of the V.L. Komarov Botanical Institute, Academy of Sciences of the USSR. Vol. 6. Equisetaceae–Butomaceae. Bibliography (Supplement 1). Compilers: V.I. Grubov, A.E. Matzenko and M.G. Pachomova. 1971. Nauka, Leningrad Division, Leningrad.

This volume is the sixth and illustrated lists of Central Asian plants (within the People's Republics of China and Mongolia) published by the Botanical Institute, Academy of Sciences of the USSR, based on the Central Asian collections of leading Russian travellers and explorers (N.M. Przewalsky, G.N. Potanin and others) as well as of Soviet expeditions, preserved in the Herbarium of the Institute.

This volume provides information on horsetails, club-mosses, conifers, joint-firs and several small families of monocotyledons standing at the beginning of the system, from Typhaceae to Butomaceae as well as a supplementary bibliographic list of works on the flora of Central Asia. In the Engler system, this volume comes immediately after the first which contained a list of ferns. Joint-firs and conifers represent the most important and interesting groups in this volume. While joint-firs fall among the main characteristic plants of wastelands, conifers represent the main species of hill forests (larch, spruce, pine and juniper).

PREFACE

This, the sixth volume of the series Plants of Central Asia, covers families Pteridophyta, Gymnospermae and up to family Butomaceae of monocotyledons in Angiospermae. Some 79 species belonging to 20 genera of 14 small families are reviewed. According to the Engler system adopted for the series, this volume should rightly have followed the first one (1963, Ferns) and preceded the fourth (1968, Gramineae).

Volume 6 also contains a list of references for the flora of Central Asia, supplementing the bibliography given in volume 1.

All monocotyledons treated here are aquatic or marshy plants, mostly with extensive geographic distribution or even cosmopolitan but of no particular interest for floristic and phytogeographic analysis. In the plant cover of Central Asia, their role is very modest as their occurrence here is rather infrequent, narrowly localised ecologically and rarely extends en masse over vast expanses.

The role of horsetails and club-mosses is altogether insignificant in the flora and plant cover of Central Asia.

Conifers, though not many (17 species), represent the main species forming montane forests of this region. *Picea schrenkiana* in Tien Shan and *P. asperata* in Nanshan and Alashan mountain range constitute the largest forest species while *Larix sibirica* is the most prominent forest species in Mongolian Atlay, Tarbagatai and Eastern Tien Shan.

Nearly all conifers are found in Central Asia only in small marginal sections of their extensive distribution ranges (*Larix sibirica, Picea asperata, P. obovata, P. schrenkiana, P. wilsonii, Pinus silvestris, P. tabulaeformis, Juniperus pseudosabina, J. sibirica* and others). Only *J. arenaria* with a very narrow distribution range in Nanshan can be regarded as endemic. Schrenk spruce with its distribution range falling almost wholly within Central Asia can be regarded as most characteristic of this region. Montane coniferous forests of Central Asia quite obviously represent highly degraded survivors of Arctic-Tertiary (Miocene–Pliocene) montane forests (forests of *Picea asperata, P. schrenkiana, P. wilsonii, Pinus tabulaeformis, Juniperus przewalskii*) as well as remnants of much later zonal taiga (forests of *Larix sibirica* and *Picea obovata*) which, in the Holocene, spread far more south than at present. All these four species of spruce belong to the same section of *Picea*, are closely interrelated and represent more autochthonous derivatives of a Tertiary species that was once extensively distributed.

Genus *Ephedra* is of utmost interest from the viewpoint of floristic history. Central Asia houses 11 species of this genus of the roughly 40 known to date. These 11 species include the narrow, highly characteristic endemic species of Alashan Gobi, *E. rhytidosperma* (section Monospermae) and subendemics *E. przewalskii* and *E. glauca* highly characteristic of Central Asia, as well as such more extensively distributed species as *E. equisetina* and *E. intermedia* which are characteristic of almost the entire eastern half of the Palaeo-Mediterranean.

Ephedra is undoubtedly a very old, Precretaceous genus and section Alatae one of its earliest branches. The distribution ranges of the two Central Asian species of this section—*E. lomatolepis* and *E. przewalskii*—deserve special attention; the former is a Kazakhstan-Middle Asian and the latter Mongolian-Kazakhstan species with isolated, evidently relict occurrence in Central Asia (Fergana, Alay mountain range, Sultan-Uizdag) and Kashmir (Hunza). Their distribution ranges overlap throughout entire Kazakhstan and Chinese Junggar plains. This feature can be interpreted as reflecting the associated formation of the desert flora of Central and Middle Asia in the early stage and of Mongolia and Junggar-Turan (Kazakhstan) province of Central Asia at a much later stage.

Two other sections of this genus—Ephedra and Monospermae—contain relatively old as well as undoubtedly recent species as judged from geographic distribution and ecological pattern. The older species are *E. glauca*, *E. intermedia* (section Ephedra) and *E. rhytidosperma*, *E. equisetina* (section Monospermae) while the latter younger ones include *E. regeliana* (section Ephedra) and *E. fedtschenkoae*, *E. monosperma* and *E. saxatilis* (section Monospermae). Phylogenetic relations and the development history of the distribution ranges of these species are quite complex and call for a special monographic treatment of the genus as a whole.

V.I. Grubov

The maps of distribution ranges in this volume were drawn by A.E. Matzenko (species of conifers) and O.I. Starikova (ephedra). Plates of plant drawings were prepared by artists G.M. Aduevskaya (Plates I–III), N.K. Voronkova (Plate IV) and E.S. Gaskevich (Plate V). O.I. Starikova also translated the Chinese texts on labels of herbarium specimens and floristic literature. For this as well as assistance in identifying the names of geographic locations, the authors express their sincere gratitude.

CONTENTS

TAXONOMY

SPECIAL ABBREVIATIONS

Abbreviations of Names of Collectors

A. Reg.	—	A. Regel
Bar.	—	V.I. Baranov
Chang	—	Y.L. Chang
Ching	—	R.C. Ching
Czet.	—	S.S. Czetyrkin
Divn.	—	D.A. Divnogorskaya
Fedcz.	—	B.A. Fedczenko (Fedtschenko)
Fet.	—	A.M. Fetisov
Glag.	—	S.A. Glagolev
Grub.	—	V.I. Grubov
Gus.	—	V.A. Gusev
Ik.-Gal.	—	N.P. Ikonnikov-Galitzkij
Ivan.	—	A.F. Ivanov
Klem.	—	E.N. Klements
Knorr.	—	O.E. Knorring
Krasch.	—	I.M. Krascheninnikov
Kryl.	—	P.N. Krylov
Kuan	—	K.C. Kuan
Lad.	—	V.F. Ladygin
Ladyzh.	—	M.V. Ladyzhensky
Lee	—	A.R. Lee
Lis.	—	V.I. Lisovsky
Litw.	—	D.I. Litwinow
Merzb.	—	G. Merzbacher
Mois.	—	V.S. Moiseenko
Pal.	—	I.V. Palibin
Pavl.	—	N.V. Pavlov
Petr.	—	M.P. Petrov
Pias.	—	P.Ya. Piassezki
Pob.	—	E.G. Pobedimova
Pop.	—	M.G. Popov

Pot.	—	G.N. Potanin
Przew.	—	N.M. Przewalsky
Rob.	—	V.I. Roborowsky
Sap.	—	V.V. Sapozhnikov
Schischk.	—	B.K. Schischkin
Serp.	—	V.M. Serpukhov
Sim. M.	—	M. Simukova
Sold.	—	V.V. Soldatov
Tug.	—	A.Ya Tugarinov
Wang	—	K.C. Wang
Yun.	—	A.A. Yunatov
Zam.	—	B.M. Zamatkinov

Abbreviations of Names of Herbaria

A	—	Arnold Arboretum: Cambridge, Massachusetts, USA
BM	—	British Museum of Natural History: London, Great Britain
FI	—	Herbarium Universitatis Florentinae, Instituto Botanico, Italy
K	—	The Herbarium, Royal Botanic Gardens: Kew, Surrey, Great Britain
Linn.	—	The Linnean Society of London: London, Great Britain

ARCHEGONIATAE–PTERIDOPHYTA

Class II. EQUISETALES

Family 5. EQUISETACEAE Rich.

1. Equisetum L.
Sp. pl. (1753) 1061; Gen. Pl. (1737) 322.

1. Stems green in winter, stiff (not flattened on drying), scabrous; perennial plants. Strobilus relatively short, acute at tip. .... 2.
+ Stems annual, rudimentary toward winter, relatively soft (flattened on drying), not scabrous. Strobilus more often oblong, obtuse. ... 4.
2. Small plant 8–25 cm tall; stems slender (less than 1 mm thick), lower part usually creeping, often curved, ascending, usually with 6 ribs; sheath tridentate. .... 7. **E. scirpoides** Michx.
+ Large plant (30) 35–75 (125) cm tall; stems thick (2–5 mm thick), with 6–30 ribs; sheath multidentate. ....3.

3. Lower part of stem with whorled branches, 2–4 mm thick; sheath 5–8 mm long, gradually enlarged upward, obconical, greyish-green, with sharply separated, rounded ribs; their teeth with deltoid whitish, sometimes even dark-coloured base with scarious margin, gradually narrowed into fairly long (up to 2 mm) and narrow scarious caducous cusp .... 6. **E. ramosissimum** Desf.

+ All stems similar, not branched, 5 mm thick, very stiff; sheath 3–10 mm long, densely appressed to stem throughout its length, usually brownish-black at base and tip; greyish-green or light-grey, more rarely monochromatic, greyish-green or black in mid-portion; teeth very short and obstuse after caducous linear-filiform cusp is shed (persistent only in uppermost sheath or young stems) ....
.... 2. **E. hiemale** L.

4. Stems of 2 types—fertile, pale or brownish in spring, differing sharply from green sterile summer stems .... 5.

+ All stems similar, green .... 7.

5. Stems withering after spore maturity; their sheaths with brownish-black, lanceolate, acute teeth numbering 8–10 (12); stems with whorled branches emerge from rhizome and succeed withered stems .... 1. **E. arvense** L.

+ Stems after spore maturity produce green branches arranged in whorls and appear similar to sterile stems manifesting simultaneously with fertile stems .... 6.

6. Sheath large (15–35 mm long) with 3–6 broad reddish-brown teeth, 2–5 joined throughout its length; branches produced on stem immediately beginning to bifurcate .... 8. **E. silvaticum** L.

+ Sheath much smaller (10–15 rarely 7 mm long) with brownish or brown narrow teeth, only some of which are joined; branches simple, not bifurcate .... 5. **E. pratense** Ehrh.

7. Stems with branches, latter intensely bifurcate, slender, long (begin to bifurcate immediately after appearance, when still undeveloped) .... 8. **E. silvaticum** L.

+ Stems simple or branched, but then branches simple, rarely only some of them whorled with few simple branchlets .... 8.

8. Plant aquatic (sometimes growing on banks of reservoirs drying up temporarily), large, 50–150 cm tall; stems fairly thick (3–6 mm thick), unbranched or with rare whorls of slender, relatively short branches; sheaths green, surrounding stem rather compactly, cylindrical, rarely barely enlarged upward, with lanceolate-subulate, brownish-black teeth along margin with very narrow whitish fringe .... 3. **E. limosum** L.

+ Plants terrestrial, much smaller (10–50 cm tall); stems 1–3 mm thick, branched .... 9.

9. Stems invariably erect, pale green (whitish or yellowish), fairly stiff; whorls of branches produced above mid-third of stem; all branches more or less equally long, not bifurcate, horizontal on stem or arcuately inclined slightly upward; sheath green, compactly adhering to stem, with narrow deltoid straight, more or less flat teeth ... . 5. **E. pratense** Ehrh.

+   Stems straight or curved; branches obliquely erect, often considerably longer in lower than upper whorls .... 10.

10. Stems dull green, often glaucescent, more or less erect; branches simple, obliquely erect, their tips somewhat curved toward stem; sheath enlarged upward, narrowly campanulate, green, with deltoid-lanceolate brownish teeth at tip and fairly broad whitish-scarious fringe .... 4. **E. palustre** L.

+   Stems bright green, straight or obliquely ascending, or arcuate in lower part and creeping later; branches obliquely erect, shorter closer to top of stem, unbranched tip of stem usually exserted under branches; sheaths on branches green, more or less longitudinally folded, as a result carinate on back, abruptly narrowing from broad base into long narrow recurved cusp .... 1. **E. arvense** L.

1. **E. arvense** L. Sp. pl. (1753) 1061; Milde, Monogr. Equiset. (1867) 218; Franch. Pl. David. 1 (1884) 343, 2 (1888) 147; Danguy in Bull. Mus. nat. hist. natur. 20 (1914) 150; Krylov, Fl. Zap. Sib. 1 (1927) 51; Iljin in Fl. SSSR, 1 (1934) 103; Hao in Engler's Bot. Jahrb. (1938) 576; Kitag. Lin. Fl. Mansh. (1939) 40; Schaffner and Li in Bull. Fan mem. Inst. Biol. (Bot.) 9, 2 (1939) 132; Norlindh, Fl. mong. steppe, 1 (1949) 35; Fl. Kirgiz. 1 (1952) 46; Grubov, Konsp. fl. MNR (1955) 54; Fl. Kazakhst. 1 (1956) 58; Fl. Tadzh. 1 (1957) 28. —**Ic.**: Milde, l.c. tab. I.

Described from Europe. Type in London (Linn.).

Banks of rivers, lakes, shoals, coastal scrubs, in floodplain and wet meadows and solonetz grasslands around springs.

IA. Mongolia: *Mong. Alt., Cis. Hing.* (Arshan town, rather dry site in wet meadow, No. 291, June 14, 1950—Chang), *Cent. Khalkha* (Ara-Dzhargalante river, Uste mountain subalp. zone, nor. slope near peak, June 10 and Aug. 12; Ubur-Dzhargalante river, meadow near river sources, Aug. 14, 30; coastal sand of Botkhona river, Sept. 5—1925, Krasch. and Zam.), *East. Mong.* (Muni-Ula, nor. slope, July 7, 1871—Przew.; Dariganga, bank of Boro-Bulak spring, Sept. 13, 1931—Pob.; "Khongkhor-obo, in valle, Aug. 11, 1926, No. 263, Eriksson"—Norlindh, l.c.; "Environs de Kailar, sables humides, alt. 750 m, No. 1440, June 22, 1896, Chaffanjon"— Danguy, l.c.), *Alash. Gobi* (Dyn'yuanin oasis, along irrigation ditches, May 31, 1908—Czet.), *Khesi* (Tszyutsyuan', rocky Gobi, May 3, 1956—Ching).

IB. Kashgar: *East.* (Bagrashkul', river valley, No. 6398, Aug. 6, 1958—A.R. Lee (1959)).

IC. Qaidam: *Plain* (Burkhan-Budda, along lower course of Nomokhun-Gol river, 3000– 3500 m, Aug. 15, 1884—Przew.).

IIA. Junggar: *Cis-Alt.* (Kran river floodplain, grass, July 8, 1959, No. 1060—A.R. Lee (1959); Ulasty river lower valley—tributary of Kran, forest glades, June 30, 1908—Sap.), *Tien Shan, Jung. Gobi* (Lower Bulugun [Grubov, l.c.]), *Zaisan* (Ch. Irtysh below Burchum, willow shoots, June 15; Belezek river lower course, tugai, June 18—1914, Schischk.), *Dzhark*. (Ili river near Kul'dzha, May 27, 1877—A. Reg.).

IIIA. Qinghai: *Nanshan* (South Tetung mountain range, mid-forest belt, June 12, 1872; same site, 2550 m, Aug. 3, 1880; south. slope of main Nanshan mountain range, Kuku-Usu river, 3000 m, July 11, 1879; Raka-Gol river, wet soil, 3000–3300 m, July 22—1880, Przew.; "Kokonor, am Fusse der gebirges Selgen, 3800 m, am Wüsten"—Hao, l.c.), *Amdo* (Baga-gorgi river, 2700 m, wet soil in forest, common, May 24–25, 1880—Przew.; Dulankhit temple, in pass through spruce forest, Aug. 9, 1901—Lad.).

IIIC. Pamir (Tashkurgan, meadow, July 25, 1913—Knorr.; King-Tau mountain range, nor. slope, upper part of forest belt, June 10, 1959—Yun.).

General distribution: Aral-Casp. Fore Balkh., Jung.-Tarb., Tien Shan; Arct., Europe, West. and East. Siberia, Mid. Asia (hilly), Far East, Nor. Mong., China (Altay, Dunbei, North, North-west, Cent., East, South-west), Himalayas (west., Kashmir), Korean peninsula, Japan, North America, nor.-west Africa and in Cape floristic region isolated in Southern Hemisphere.

2. **E. hiemale** L. Sp. pl. (1753) 1062; Milde, Monogr. Equiset. (1867) 511; Krylov, Fl. Zap. Sib. 1 (1927) 57; Iljin in Fl. SSSR, 1 (1934) 110; Kitag. Lin. Fl. Mansh. (1939) 41; Schaffner and Li in Bull. Fan mem. Inst. Biol. (Bot.) 9, 2 (1939) 130; Fl. Kirgiz. 1 (1952) 45; Fl. Kazakhst. 1 (1956) 60; Hauke in Amer. Fern J. 52, 2 (1962) 58. —**Ic.**: Milde, l.c. tab. XXIX.

Described from Europe. Type in London (Linn.).

Coniferous and sometimes birch forests, scrubs, banks of forest rivers, ravines and irrigation ditches.

IIA. Junggar: *Tien Shan* (Sary-Chagan [June] 1875—Larionov; Kokkamyr, Sarybulak, 1800–2100 m, April 29, 1878; Borgaty, 1500–1800 m, July 5; Aryslyn esturay, 2400 m, July 10; Aryslyn, 2400 m, July 14; Aryslyn estuary, 1800 m, July 20; near confluence of Kash and Aryslyn, 1650 m, July 25—1879, A. Reg.; Kelisu, fringes of irrigation ditches, No. 1876, July 17; 20 km nor. of Ulastai, slope, No. 3865, Aug. 28; Tsitai [Guchen], Magolyan, rubble, No. 4411, Sept. 22—1957, Kuan), *Zaisan* (Blandy-Kul'-kum sand, right bank of Alkabek, July 6, 1900—Reznichenko; Alkabek, sand, Aug. 6, 1908—Fedtsch.).

General distribution: Aral-Casp., Fore Balkh., Jung.-Tarb., Nor. and Cent. Tien Shan; Arct. (Europ.), Caucasus, forest zone of Europe and Asia, China (Altay, Dunbei), Korean peninsula, Japan and North America (forest zone).

3. **E. limosum** L. Sp. pl. (1753) 1062; Milde, Monogr. Equiset. (1967) 339; Kitag. Lin. Fl. Mansh. (1939) 41. —*E. heleocharis* Ehrh. Hannov. Mag. 9 (1784) 286; Krylov, Fl. Zap. Sib. 1 (1927) 55; Iljin in Fl. SSSR, 1 (1934) 108; Grubov, Konsp. fl. MNR (1955) 54; Fl. Kazakhst. 1 (1956) 60. —*E. fluviatile* L. Sp. pl. (1753) 1062; Schaffner and Li in Bull. Fan mem. Inst. Biol. (Bot.) 9, 2 (1939) 131.—**Ic.**: Milde, l.c. tab. XV.

Described from Europe. Type in London (Linn.).

Lakes, backwaters, marshes, irrigation ditches, floodplain meadows, usually in water or sometimes in temporarily drying up places.

IA. Mongolia: *Bas. Lakes* (Ulangom, marsh, June 13, 1924—Neiburg).

IB. Kashgar: *East.* (Toksun district, river bank, 500 m, No. 7281, June 15, 1958—A.R. Lee (1959)).

IIA. Junggar: *Cis-Alt.* (Qinhe [Chingil'] No. 1648, Aug. 11; west of Fuyun' town [Koktogoi] No. 1890, Aug. 13; same site, No. 2355, Aug. 25—1956, Ching).

General distribution: Fore Balkh.; Arct., Europe, Mediterr., Balk.-Asia Minor, Caucasus, West. and East. Sib., Far East, Nor. Mongolia (Fore Hubs., Hent., Mong.-Daur.), China (Dunbei), Japan, North America (forest zone).

4. **E. palustre** L. Sp. pl. (1753) 1061; Milde, Monogr. Equiset. (1867) 323; Krylov, Fl. Zap. Sib. 1 (1927) 54; Iljin in Fl. SSSR, 1 (1934) 108; Kitag. Lin. Fl. Mansh. (1939) 41; Schaffner and Li in Bull. Fan mem. Inst. Biol. (Bot.) 9, 2 (1939) 131; Fl. Kirgiz. 1 (1952) 46; Grubov, Konsp. fl. MNR (1955) 55; Fl. Kazakhst. 1 (1956) 59. —Ic.: Milde, l.c. tab. XIII.

Described from Europe. Type in London (Linn.).

Grassy and mossy marshes, wet and marshy banks of rivers and lakes, wet meadows.

IA. **Mongolia:** *East. Mong.* (Khailar, near water or on bank, No. 558, Aug. 6, 1951 —S.H. Li et al. (1951)), *Bas. Lakes* (Ulangom, marsh, June 2, 1879 —Pot.).

IIA. **Junggar:** *Tien Shan* (Piluchi, 900–1500 m, April 24, 1879—A. Reg.).

General distribution: Aral-Casp., Fore Balkh., Jung.-Tarb.; Arct., Europe, Mediterr., Balk.-Asia Minor, Causasus, West. Sib., Far East, Nor. Mong., China (Dunbei, North-west), Korean peninsula, Japan, North America (forest zone).

5. **E. pratense** Ehrh. Hannov. Mag. 9 (1784) 138; Milde, Monogr. Equiset. (1867) 263; Krylov, Fl. Zap. Sib. 1 (1927) 52; Iljin in Fl. SSSR, 1 (1934) 104; Kitag. Lin. Fl. Mansh. (1939) 41; Schaffner and Li in Bull. Fan mem. Inst. Biol. (Bot.) 9, 2 (1939) 132; Fl. Kirgiz. 1 (1952) 49; Grubov, Konsp. fl. MNR (1955) 55; Fl. Kazakhst. 1 (1956) 509. —*E. umbrosum* J.G.F. Mey. in Willd. Enum. pl. horti Berol. (1809) 1065. —Ic.: Milde, l.c. tab. VII.

Described from Europe. Type in Geneva (G-DC).

Dry, more rarely wet meadows, thin forests, forest clearing, forest fringes, as well as banks of rivers and lakes, scrubs and bottom-land deciduous forests.

IA. **Mongolia:** *Cis-Hing.* (Yakshi railway station, June 11, 1902—Litw.; Arshan, wet meadow, No. 292, June 13, 1950—Chang), *East. Mong.* (nor. slope of Muni-Ula mountain range, wet forest, June 25, 1871—Przew.).

IIA. **Junggar:** *Tien Shan* (nor. slope, Tsitai [Guchen], Magolyan, along irrigation ditches, No. 4390, Sept. 22; between Daban and Danyu, on slope, July 19—1957, Kuan).

General distribution: Fore Balkh.; Arct., Mid. Asia (Chatkal), Europe, West. and East. Sib., Far East, Nor. Mong. (Hent., Hang., Mong.-Daur.), China (Dunbei), Nor. America (forest zone).

6. **E. ramosissimum** Desf. Fl. Atl. 2 (1799) 398; Milde, Monogr. Equiset. (1867) 428; Franch. Pl. David. 1 (1884) 343; Danguy in Bull. Mus. nat. hist. natur. 17. (1911) 413, 20 (1914) 150; Christ. in J. Wash. Ac. Sci. 17 (1927) 499; Krylov, Fl. Zap. Sib. 1 (1927) 56; Pampanini, Fl. Carac. (1930) 67; Iljin in Fl. SSSR, 1 (1934) 109; Persson in Bot. notiser (1938) 272; Kitag. Lin. Fl. Mansh. (1939) 41; Schaffner and Li in Bull. Fan mem. Inst. Biol. (Bot.) 9, 2 (1939) 129; Norlindh, Fl. mong. steppe, 1 (1949) 35; Fl. Kirgiz. 1 (1952) 46; Grubov, Konsp. fl. MNR (1955) 55; Fl. Kazakhst. 1 (1956) 60; Fl. Tadzh. 1 (1957) 28;

Chen and Chou, Rast. pokrov r. Sulekhe (1957) 86. —*E. ramosissimum* subsp. *ramosissimum* Hauke in Amer. Fern J. 52, 1 (1962) 32. —*E. ramosum* DC. Syn. Pl. Fl. Gall. (1806) 118. —*E. multicaule* Ledeb. Fl. Ross. 4 (1853) 490. —Ic.: Milde, l.c. tab. XXIV; Norlindh, l.c. fig. 2.

Described from Nor. Africa. Type in Paris. Plate I.

Puffed solonchaks, solonetz sand, chalk exposures and along precipices, river valleys and shoals.

IA. Mongolia: *Mong. Alt.* (Bulugun region, the Dzhirgalante river, puffed sandy solonchaks, Sept. 16, 1930—Bar.), *East. Mong.* (right bank of Huang He river around Hekou town, rocky and sandy soil, Aug. 4, 9, 10—1884, Pot.), *Alash. Gobi, Khesi.*

IB. Kashgar: *Nor.* (Yarkend-Darya, along channels, on wet alluvial clay, 900 m, June 21, 1889—Rob.; Uchturfan, bank of spring, May 27 and June 9, 1908—Divn.; 60 km south from Shakh'yar, old river bed, No. 8482, Sept. 26, 1958 —Lee and Chu (A.R. Lee (1959)), *West.* (Yangissar environs, bank of irrigation ditch, sand, May 26, 1909—Divn.; Artush, in maize crop, No. 7504, Sept. 11, 1958—A.R. Lee (1959); "Bostan-terek, ca. 2400 m, Aug. 14, 1934; Jerzil, 2800 m, July 23, 1930"—Persson, l.c.), *South.* (nor. slope of Russky mountain range, along Mal'dzha river, among shrubs, 2160 m, May 16, 1885—Przew.; Russky mountain range, Mal'dzha river, among shrubs, 2100 m, June 16, 1890—Rob.; Keriya river valley near intersection with road at Niyu, floodplain, silted meander, solonchak meadow, May 9; Sandzhu oasis, along irrigation ditches, May 27—1959, Yun.; Sandzhu-Bazar, in farm, No. 161, May 28, 1959—A.R. Lee (1959)), *East.* (Khami, May 15–16, 1879—Przew.; in Khami region, near water, No. 445, May 4, 1957—Kuan; Turfan, Putagou, along brim of irrigation canal, No. 5514, June 1; nor.-east of Toksun town, Syaotsaokhu lake, in water, No. 7314, June 19—1958, Lee and Chu (A.R. Lee (1959)); Kurlya, No. 6841, July 20; Yan'tszy [Karashar] in Bagrashkul' lake region, No. 845, July 24; in Bachu [Maralbashi] region, No. 7520, Sept. 14—1958, A.R. Lee (1959)).

IIA. Junggar: *Cis-Alt.* (in Qinhe [Chingil'] region, in river gorge, 2000 m, No. 1542, Aug. 8, 1956—Ching), *Tien Shan, Jung. Gobi, Zaisan* (left bank of Ch. Irtysh westward opposite Cherektas hill, tugai, June 11; Kaba river near Kaba village, tugai, June 16; Belezeka river lower course, tugai, June 18—1914, Schischk.), *Dzhark.* (Ili river near Kul'dzha, May 3, 6, 8, 29 and 30, 1877—A.Reg.).

IIIA. Qinghai: *Amdo* (upper Huang He, 2400 m, June 2, 1880—Przew.).

IIIC. Pamir (between Pishtan and Kara-chukur, in Dagnyn-bash valley, meadows, 3600 m, July 15, 1909—Alekseenko, Pakhpu river gorge, 2700 m, Aug. 2, 1942—Serp.; Tashkurgan, in valley, 3000 m, No. 294, June 13, 1959—A.R. Lee (1959)).

General distribution: Aral-Casp., Fore Balkh., Jung.-Tarb., Nor. and Cent. Tien Shan, East. Pamir; Europe (south.), Mediterr., Balk.-Asia Minor, Fore Asia, Caucasus, Mid. Asia, West. Sib. (Altay), China (Dunbei, North, North-west, Cent., East, South-west., South, Hainan, Taiwan), Himalayas, Japan, North and South America, Nor. and South Africa.

7. **E. scirpoides** Michx. Fl. Bor.-Am. 2 (1803) 281; Milde, Monogr. Equiset. (1867) 596; Krylov, Fl. Zap. Sib. 1 (1927) 58; Iljin in Fl. SSSR, 1 (1934) 111; Grubov, Konsp. fl. MNR (1955) 55; Fl. Kazakhst. 1 (1956) 61; Hauke in Amer. Fern J. 52, 2 (1962) 63. —Ic.: Milde, l.c. tab. XXXV.

Described from North America. Type in Paris.

Mossy swamps, along river banks, swampy forests, usually on moss.

IIA. Junggar: *Tien Shan* (east of Shichan, spruce forest, scrubs, slopes, No. 3549, Oct. 3, 1956—Ching).

General distribution: Arct., Europe (nor.), West. and East. Sib., Far East, Nor. Mong. (Hent., Mong.-Daur.), China (Dunbei), North America.

8. **E. silvaticum** L. Sp. pl. (1753) 1061; Milde, Monogr. Equiset. (1867) 287; Krylov, Fl. Zap. Sib. 1 (1927) 53; Iljin in Fl. SSSR, 1 (1934) 107; Kitag. Lin. Fl. Mansh. (1939) 41; Schaffner and Li in Bull. Fan mem. Inst. Biol. (Bot.) 9, 2 (1939) 132; Grubov, Konsp. fl. MNR (1955) 55; Fl. Kazakhst. 1 (1956) 59. —Ic.: Milde, l.c. tab. IX.

Described from Nor. Europe. Type in London (Linn.).

Forests, more often coniferous, forest meadows, scrubs.

IA. **Mongolia:** *Cis-Hing.* (Arshan town, mountain meadow, No. 290, June 14, 1950—Chang).

General distribution: Aral-Casp. (nor.), Fore Balkh., Jung.-Tarb.; Arct., Europe, Balk.-Asia Minor, Caucasus, Mid. Asia, West. Sib., East. Sib., Far East, Nor. Mongolia (Fore Hubs., Hent.), China (Dunbei), Korean peninsula, Japan, North America.

Class III. *LYCOPODIALES*

1. Leaves (phylloids) without ligules, spirally arranged, appearing decussate in dorsiventral forms. All spores similar. Evergreen plants with decumbent or erect stem .... 6. **Lycopodiaceae** Reichb.
+ Leaves (phylloids) with ligules arranged alternately or in 4 oblong rows. Spores of 2 types. Small, moss-like plants forming more or less compact mats .... 7. **Selaginellaceae** Mett.

Family 6. **LYCOPODIACEAE** Reichb.

1. **Lycopodium** L.
Sp. pl. (1753) 1100

1. **L. selago** L. Sp. pl. (1753) 1102; Krylov, Fl. Zap. Sib. 1 (1927) 60; Iljin in Fl. SSSR, 1 (1934) 114; Grubov, Konsp. fl. MNR (1955) 55; Fl. Kazakhst. 1 (1956) 62. —Ic.: Fl. SSSR, 1, Plate VI, fig. 1a–c.

Described from Nor. Europe. Type in London (Linn.).

Mountain boreal forests, along upper forest boundary, meadow slopes in alpine belt.

IIA. **Junggar:** *Tien Shan* (Aryslyn, 2400 m, July 12 and 17; Kunges, Aug. 27—1879, A. Reg.).

General distribution: Subcosmopolitan.

Family 7. **SELAGINELLACEAE** Mett.

1. **Selaginella** Beauv.
Prodr. (1805) 101.

1. Stems brown; all branches flat, invariably distinctly dorsiventral
.... 2.

+ Stems red or bright orange; all branches equilateral or dorsiventral or equilateral only in upper part of plant .... 3.

2. Stems creeping, rooting; branches not compact, with considerable gaps (interrupted); leaves not convoluted along margin (in herbarium) .... 3. **S. sinensis** (Desv.) Spring.

+ Stems in clusters (rosettes); leaves convoluted along margin (in herbarium) .... 4. **S. tamariscina** (Beauv.) Spring.

3. Dorsiventral branches extremely rare; leaves acutely carinate, acuminate, resembling short spine, entire ....
.... 2. **S. sanguinolenta** (L.) Spring.

+ Dorsiventral branches nearly always present; leaves convex on back, with obtuse keel, obtuse or barely subacute, with short cilia along margin .... 1. **S. borealis** (Kaulf.) Rupr.

1. **S. borealis** (Kaulf.) Rupr. in Beitr. Pflanzenk. Russ. Reich. 3 (1845) 32; Spring, Monogr. Lycopod. 2 (1848) 96; Ledeb. Fl. Ross. 4 (1853) 502; Hieron in Engler-Prantl, Pflanzenfam. 1, 4 (1900) 674; Iljin in Fl. SSSR, 1 (1934) 125; Grubov, Konsp. fl. MNR (1955) 56. —*S. sanguinolenta* auct.: Alston in Bull. Fan mem. Inst. Biol. (Bot.) 5, 6 (1934) 267. —*Lycopodium boreale* Kaulf. Enum. filic. (1824) 17.

Described from Kamchatka. Type in Berlin.

Rock crevices and among rocks.

IA. Mongolia: *Cent. Khalkha* (rock crevices on Dzakha mountain in Suchzhiin-Gol river valley, July 10, 1924—Pavl.; environs of Ikhe-Tukhum-nur lake, Ongon-Khairkhan mountain, July 28, 1926—Zam.; rocky slope of Khairkhan mountain, Sept. 27, 1931—Ik.-Gal.; Bain-Ula hill [Sorgol-Khairkhan], granite rocks on south. slope, July 12, 1948—Grub.).

IIA. Junggar: *Tien Shan* (Kapsalan pass, 3600–3900 m, June 16, 1876—A. Reg.).

General distribution: East. Sib. (Ang.-Sayan., Daur., Leno-Kolym.), Far East, Nor. Mong. (Hang.), China (Dunbei, North, North-west), Japan (?).

Note. It is difficult to discern any stable characteristics that differentiate *S. sanguinolenta* (L.) Spring from *S. borealis* (Kaulf.) Rupr. Along with extreme, well-distinguished races, many intergrades are seen. According to some investigators (Goebel and Suessenguth, Flora, 122: 393–402), heterophyllous races resulting from excessive moisture can be experimentally produced. The distribution ranges of both species are nearly similar. Evidently, Alston (l.c.) was right, merging the two.

2. **S. sanguinolenta** (L.) Spring, Monogr. Lycopod. 2 (1848) 57; Ledeb. Fl. Ross. 4 (1853) 501; Franch. Pl. David. 1 (1884) 343; Hieron in Engler-Prantl, Pflanzenfam. 1, 4 (1900) 673; Krylov, Fl. Zap. Sib. 1 (1927) 68; Iljin in Fl. SSSR, 1 (1934) 125; Alston in Bull. Fan mem. Inst. Biol (Bot.) 5, 6 (1934) 267; Kitag. Lin. Fl. Mansh. (1939) 44; Grubov, Konsp. fl. MNR (1955) 56. —*Lycopodium sanguinolenta* L. Sp. pl. (1753) 1104.

Described from Kamchatka. Type in London (Linn.).

Rock crevices, among stones.

IA. Mongolia: *Cent. Khalkha* (Dzakha hill in Suchzhiin-Gol river valley, rock crevices, July 9, 1924—Pavl.; Orochen-Sume environs of monastery, granite exposures, Aug. 23, 1925—

Krasch. and Zam.; Buridu somon, 15–20 km from somon, Ikhe-Mongol-obo-ula hill, upper part of rocky slope, June 15, 1952—Davazamč, *Alash. Gobi* (Bayan-Khoto—Inchuan' road, gorge in Alashan moutain range, rocks, under shade, common, July 7, 1957—Kabanov; 50 km south-west of Inchuan' town, Alashan mountain range, rocky slopes, hilly semi-desert, July 10, 1957—Petr.).

IB. Kashgar: *Nor.* (Muzart river gorge, Aug. [15–17], 1877—A. Reg.).

IIA. Junggar: *Tien Shan* (Akburtash mountains, 1200–1500 m, July 22; Sharysu river, 2100–2400 m, July 25—1878, A. Reg.; Ketmen' mountain range, 8–10 km beyond Sarbushin settlement, steppe belt, south. rocky slope, Aug. 23, 1957—Yun.).

General distribution: East. Sib. (Ang.-Sayan., Daur., Leno-Kol.), Far East, Nor. Mongolia (Hang.), China (Dunbei, North, North-west, Central), Japan (?).

3. **S. sinensis** (Desv.) Spring, Monogr. Lycopod. 2 (1848) 75; Alston in Bull. Fan mem. Inst. Biol. (Bot.) 5, 6 (1934) 269; Kitag. Lin. Fl. Mansh. (1939) 44. —*S. mongholica* Rupr. Beitr. Pflanzenk. Russ. Reich. 3 (1845) 32; Spring, l.c. 262; Franch. Pl. David. 1 (1884) 343; Hieron in Engler-Prantl, Pflanzenfam. 1, 4 (1900) 674. —*Lycopodium sinense* Desv. in Ann. Soc. Linn. Paris, 6 (1827) 189.

Described from North China. Type in Paris.

On rocks.

IA. Mongolia: *East. Mong.* (between Shara-Muren river and Tun'tszyaintszy, on right bank of Shara-Muren river, near mountain base, 1899—Pal.), *Khesi* (Sachzhou [Dun'khuan] oasis, on Daikhe river, Sept. 11, 1890—Marten).

General distribution: China (Dunbei, North, North-west, Centre, East).

4. **S. tamariscina** (Beauv.) Spring in Bull. Ac. Brux. (1843) 136; Hand.-Mazz. Symb. Sin. 5 (1929) 5; Alston in Bull. Fan mem. Inst. Biol. (Bot.) 5, 6 (1934) 270. —*S. involvens* (Sw.) Spring, l.c. 136; Franch. Pl. David. 1 (1884) 345; Danguy in Bull. Mus. nat. hist. natur. 20 (1914) 150; Iljin in Fl. SSSR, 1 (1934) 126; Kitag. Lin. Fl. Mansh. (1939) 43. —*S. pulvinata* (Hook. et Grev.) Maxim. in Mém. Ac. Sci. St.-Pétersb. 9 (1859) 335. —*S. tamariscina* var. *pulvinata* Alston, l.c. 271. —*Stachygynandrum tamariscinum* Beauv. Prodr. (1805) 106. —*Lycopodium involvens* Sw. Syn. filic. (1806) 182. —*L. tamariscinum* (Beauv.) Desv. in Poir. Encycl. Suppl. 3 (1814) 540. —*L. pulvinatum* Hook. et Grev. in Hook. Bot. Misc. 2 (1831) 381.

Described from East. India. Type in Berlin.

On rocks.

IIIB. Tibet: *South.* ("Lhasa, 3600 m, Waddel"—Alston, l.c. 271).

General distribution: Far East (south), China (all regions except Altay and Hainan), Himalayas (west.), India (nor.), Korean peninsula, Japan.

EMBRYOPHYTA–SIPHONOGAMA

Subdivision I. GYMNOSPERMAE

1. Trees or rarely large shrubs. Strobili unisexual; plants monoecious
   .... Class I. **Coniferales.**

+ Xerophilous or semi-xerophilous highly branched shrubs. Strobili unisexual; plants generally dioecious .... Class II. **Gnetales**.

Class I. CONIFERALES

1. All leaves (needles) arranged spirally on long shoots and in clusters on short ones. Cones with several woody ovuliferous scales arranged in axils of coriaceous bract scales. Seeds more or less winged .... 8. **Pinaceae** Lindl.
+ Leaves acicular, in whorls of 3 or leaves scale-like, opposite. Ovuliferous scales usually 3, bract scales lacking; cones baccate; ovuliferous scales becoming fleshy and completely coalesce after fertilisation .... 9. **Cupressaceae** Bartl.

Family 8. PINACEAE Lindl.

1. Shoots of same type, invariably long. Cones wholly deciduous. Leaf scars with elevated, rather oblong, cushions .... 1. **Picea** Dietr.
+ Shoots of 2 types, long and short .... 2.
2. Leaves deciduous in winter. Cones small, up to 3 cm long, maturing in 1 year .... 2. **Larix** Mill.
+ Leaves persistent in winter. Cones very large, maturing in 2–3 years .... 3. **Pinus** L.

1. Picea A. Dietr.
Fl. Berlin (1824) 479.

1. Small tree up to 11 m tall. Cones small (4–6 cm long), hanging together in small numbers; ovuliferous scales erose at tip. Needles up to 10 mm long; free part of pulvinus emerging from shoot at very acute angle as though adhering to shoot; young shoots glabrous ....
.... 4. **P. wilsonii** Mast.
+ Trees up to 30–40 m tall. Cones very large (5–16 cm long), singly or few together; ovuliferous scales rounded at tip, entire. Needles up to 35 mm long .... 2.
2. Free part of pulvinus emerging from shoot at acute or very acute angle; young shoots more or less pubescent. Needles 10–20 mm long. Cones 5–8 cm long .... 2. **P. obovata** Ledeb.
+ Free part of pulvinus emerging from shoot at nearly right or obtuse angle. Cones 7–16 cm long .... 3.
3. Attached part of pulvinus highly convex, ectypal; its free part emerging from shoot at right angle or subobtuse angle; young shoots more or less pubescent, often with glaucescent bloom. Needles 12–15 mm long. Cones 8–12 cm long .... 1. **P. asperata** Mast.

+   Attached part of pulvinus poorly convex; its free part emerging from shoot at acute or nearly right angle; young shoots glabrous. Needles 20–35 mm long. Cones 7–16 cm long ....
.... 3. **P. schrenkiana** Fisch. et Mey.

1. **P. asperata** Mast. in J. Linn. Soc. London (Bot.) 37 (1906) 419; Rehder and Wilson in Sargent, Pl. Wilson. 2 (1914) 22; id. in J. Arn. Arb. 9 (1928) 8; Ching in Bull. Fan mem. Inst. Biol. (Bot.) 10, 5 (1941) 258; Walker in Contribs U.S. Nat. Herb. 28 (1941) 593; Dallimore and Jackson, Handb. Conif. (1964) 343. —*P. crassifolia* Kom. in Bot. mat. (Leningrad) 4 (1923) 177. —**Ic.:** Dallimore and Jackson, l.c. fig. 66.

Described from China (Sichuan). Type in Arnold Arboretum (A). Map. 1.

Forms montane forests, pure and mixed, 2100–3600 m alt.; most common from 3000 to 3300m alt.

IA. Mongolia: *Alash. Gobi* (Alashan, midbelt, forest, June 18, 1884—Przew.; Tsuburgin-Gol gorge, April 28, 1908; Dyn'yuan'in [Bain-Khoto], in garden, April 14, 1909 —Czet.; same site, 30 km east of town, Baisy monastery, spruce-pine forest, 2500 m, July 6, 1957—Petr.; "Ho-Lan-Shan", Ching, l.c.).

IC. Qaidam: *mountains* (in Karagainyn-Gol river valley, Aug. 10, 1893—Rob.; Karagainyn-Ula mountains, Karagainyn-Gol river valley, 3300–3600 m, May 17, 1895—Kozlov).

IIIA. Qinghai: *Nanshan, Amdo.*

IIIB. Tibet: *Weitzan* ("Amnyi-Machen range, Hjachen valley, near Yellow River, No. 14444, July 16, 1926, Rock"—Rehder and Wilson, l.c. [1928]).

General distribution: China (North-west, South-west, Sichuan).

Note. This species of spruce is distributed in the Alashan mountain range but *P. schrenkiana* Fisch. et Mey. is also found occasionally in the same region. The 2 obviously hybridise in the area of contact of their distribution ranges. Some such hybrids are preserved in the Herbarium Botanic Institute.

We could detect no feature that could differentiate *P. crassifolia* Kom. from *P. asperta* Mast. and hence have placed it among synonyms of the latter.

2. **P. obovata** Ledeb. Fl. Alt. 4 (1833) 201; Simpson in J. Linn. Soc. London (Bot.) 41 (1912–1913) 444; Krylov, Fl. Zap. Sib. 1 (1927) 73; Komarov in Fl. SSSR, 1 (1934) 145; Kitag. Lin. Fl. Mansh. (1939) 47; Grubov, Konsp. fl. MNR (1955) 56; Fl. Kazakhst. 1 (1956) 66; Dallimore and Jackson, Handb. Conif. (1964) 366. —**Ic.:** Ledeb. Ic. Pl. fl. ross. 5 (1833) tab. 499.

Described from Altay. Type in Leningrad.

Forms taiga forests in mountains; found along river banks, talus and placers, 800–2000 m alt.

IIA. Junggar: *Cis.-Alt.* (in Kandagatai river valley, Sept. 13, 1876—Pot.; Durul'chen river valley, June 29, 1906 —Sap.; south of Fuiyun' [Koktogoi], Ukagou village, 1200 m, No. 1792, Aug. 11; Shara-Sume region, 1800 m, No. 2557, Aug. 27—1956, Ching; Altay hills, valley slope under shade, 1500 m, No. 10722, July 21, 1959—A.R. Lee (1959); "Great Altai Mts. 900 to 1950 m, Kran River, Price"—Simpson, l.c.).

General distribution: Arct., Europe (east. Europ. part of USSR), West. and East. Sib., Far East (nor. Amur region), Nor. Mong., China (Dunbei).

3. **P. schrenkiana** Fisch. et Mey. Bull. Ac. Sci. St.-Pétersb. 10 (1842) 253; Forbes and Hemsley, Index Fl. Sin. 2 (1902) 554; Mast. in J. Linn. Soc. London (Bot.) 37 (1906) 419; Simpson in J. Linn. Soc. London (Bot.) 41 (1912–1913) 444; Rehder and Wilson in Sargent, Pl. Wils. 2 (1914) 29, p.p.; Komarov in Fl. SSSR, 1 (1934) 147; Persson in Bot. notiser (1938) 272; Fl. Kirgiz. 1 (1952) 51; Fl. Kazakhst. 1 (1956) 66; Dallimore and Jackson, Handb. Conif. (1964) 376. —*P. tianschanica* Rupr. in Mém. Ac. Sci. St.-Pétersb. 7 ser. 14, 4 (1869) 72; Komarov in Fl. SSSR, 1 (1934) 147. —*P. robertii* P. Wipp. in Dokl. AN SSSR, 61, 2 (1948) 373; Fl. Kirgiz. 1 (1952) 51. —**Ic.:** Fl. SSSR, 1, Plate VII, fig. 18 and Plate VIII, fig. 6; Fl. Kazakhst. 1 Plate III, fig. 3.

Described from Tien Shan. Type in Leningrad. Map 1.

Forms pure and mixed forests on steep northern slopes in mountains, in deep creek valleys and gorges, often on rocks and talus, 800 (1100) to 3000 m alt.; most common at 2500 m alt.

IA. **Mongolia:** (*Alash. Gobi* (Alashan mountain range, Dyn'yuanin [Bayan-Khoto] oasis, in garden, Aug. 17, 1909—Czet.; environs of Bayan-Khoto town, spur of Alashan mountain range, in narrow ravines under shade, July 6, 1957—Kabanov).

IB. **Kashgar:** *Nor.* (Kurdai, Karabulak, July 29, 1907—Merzbacher; Bogno pass along Charlung river, 3030–3500 m, July 29, 1941— Serp.; Muzart river valley, Tupu-Daban, near Oi-Terek area, 2900 m, Sept. 7; same site, 5–6 km west of Kurgai settlement, grassy spruce grove, Sept. 12—1958, Yun.; Bai district, slope, No. 8246, Sept. 7; Aksu district, along Muzart river, Pochentszy settlement, slope, 2100 m, No. 8334, Sept. 12, 1958—Lee and Chu (A.R. Lee (1959)); Kucha-Shakh'yar, 2750 m, No. 10068, July 26, 1959—A.R. Lee (1959)); *West.* (nor. slope of Tokhta-Khon hill, July 21, 1889—Rob.; King-Tau mountain range, nor. slope, 4 km south-east of Kosh-Kulak settlement, forest belt, June 10, 1959—Yun.; Kosh-Kulak—Upal, on nor. slope, 2500 m, No. 229; same site, 3000 m, No. 239—June 10, 1959, A.R. Lee (1959); Irkeshtam).

IIA. **Junggar:** *Jung. Alt.* (in Toli region, No. 1118, 2509, Aug. 6; same site, No. 1355, Aug. 7—1958, Kuan), *Tien Shan.*

General distribution: Jung.-Tarb., Nor. and Cent. Tien Shan; Mid. Asia (Pamiro-Alay: Dzhusaly river).

Note. Most widely distributed species of spruce in Cent. Asia forming forests over large expanses on Junggar and Kashgar mountains. Some specimens of this species have been found in Alashan mountain range but some were probably hybrids (*P. schrenkiana* × *P. asperata*).

4. **P. wilsonii** Mast. in Gard. Chron. ser. 3, 37 (1903) 133; Rehder and Wilson in Sargent, Pl. Wilson, 2 (1914) 27; id. in J. Arn. Arb. 9 (1928) 10; Hao in Engler's Bot. Jahrb. 68 (1938) 578; Walker in Contribs U.S. Nat. Herb. 28 (1941) 593; Dallimore and Jackson, Handb. Conif. (1964) 382.—*P. watsoniana* Mast. in J. Linn. Soc. London (Bot.) 37 (1906) 419; Rehder and Wilson, l.c. (1914) 27. —**Ic.:** Gard. Chron. 37 (1903) fig. 55, 56; Dallimore and Jackson, l.c. fig. 79.

Described from Cent. China (Hubei). Type in Arnold Arboretum (A). Forms forests at 2400–3200 m alt.

IIIA. **Qinghai:** *Nanshan* (South Tetung mountain range, forest belt, nor. slope, Aug. 6, 1872; Tetung river valley, between Rangkhta-Gol estuary and Chertynton temple, 2250 m, Aug. 7, 1880; Tetung river valley, near Chertynton temple, March 4, 1884—Przew.; Cheibsen temple, 2100–3000 m, Sept. 1901, Lad.; "Ju Er Ping, Tai Wang-Kou, in *Picea* forests up to 3000 m, Ching"—Walker, l.c.), *Amdo* (Dulan-khit temple, 3300–3600 m, Aug. 9, 1901—Lad.).

General distribution: China (North, North-west, Cent., South-west).

Note. Most widely distributed species of spruce in China. *P. watsoniana* Mast. described by the same author somewhat later than *P. wilsonii* Mast. does not differ: complete identity of characteristics and distribution range is striking and hence we place it among synonyms.

2. Larix Mill.
Gard. Dict. ed. 7 (1759).

1. Mature cones more or less ovate, 2–3 cm long; ovuliferous scales broadly ovate or suborbicular, with more or less rounded margin, slightly incurved (spoon-shaped) at tip; scales generally exceeding 20 .... 1. **L. sibirica** Ledeb.

+ Mature cones invariably widely divaricate, not more than 2 cm long (usually 1.5 cm); ovuliferous scales spatulate with straight abscised or sinuate margin, not incurved; scales less than 20 ....
.... **L. dahurica** Turcz.

1. **L. sibirica** Ledeb. Fl. Alt. 4 (1833) 204; Franch. Pl. David. 1 (1884) 287; Forbes and Hemsley, Index Fl. Sin. 2 (1902) 558; Simpson in J. Linn. Soc. London (Bot.) 41 (1912–1913) 445; Danguy in Bull. Mus. nat. hist. natur. 20 (1914) 149; Krylov, Fl. Zap. Sib. 1 (1927) 75; Komarov in Fl. SSSR, 1 (1934) 155; Fl. Kirgiz. 1 (1952) 55; Grubov, Konsp. fl. MNR (1955) 56; Fl. Kazakhst. 1 (1956) 68. —Ic.: Fl. SSSR, 1, Plate VII, figs, 1, 2, 15, Plate VIII, fig. 9, Fl. Kazakhst. 1, Plate III, fig. 5.

Described from Urals. Type in Leningrad. Map 2.

Main forest-forming species of hill forests, sometimes along river valleys in lower mountain belt.

IA. **Mongolia:** *Khobd.* (Khatu river, Bukhu-Muren tributary, Aug. 7, 1909—Sap.), *Mong. Alt.* (Urtu-Gola valley, larch forest, on east. slope, Aug. 17 and 19; Khara-Dzarga mountain range, Khairkhan-Duru environs, larch forest, Aug. 26; Shutyn-Gol river valley, larch forest on nor. slope, Aug. 29—1930, Pob.; Khan-Taishiri mountain range, nor. slope, Khalyun area, larch forest, Aug. 24, 1943; Khan-Taishiri mountain range, nor. slope, July 11 and 12, 1945; same site, larch forest, July 14; in upper course of Dundu-Tumurte river [Bulugun tributary], along brook, July 25, 1947—Yun.), *Bas. Lakes* (in Shuryk river valley [Khirgis-Nora system], July 23, 1877; Ulan-Daban pass [Ubsa lake], June 27; river bank in Ubsa lake basin, 900 m, July 3; in Ulangom valley, Sept. 6 —1879, Pot.; larch-birch forest, in valley, 15 km nor.-nor.-west of Ulangom town, July 29, 1945—Yun.).

IIA. **Junggar:** *Cis.Alt., Tarb.* (in Khobuk-Saira region, in mountain gorge, No. 3418, Sept. 23, 1956—Ching; same site, Chagan-obo mountain, 2000 m, No. 10541, June 26, 1959—A.R. Lee (1959); Saur mountain range, southern slope, Karagaitu river gorge, Bain-Tsagan creek valley, grassy larch grove, nor. slope, June 23, 1957—Yun.), *Tien Shan* (between Barkul lake and Khami, May 22, 1877—Przew.; Koshety-Daban pass, nor. of Khami town, May 23, 1877—Pot.; north of Barkul' lake, Sept. 20, 1895—Rob.; in Barkul' lake region, nor. slope, Nos. 2159, 4927, Sept. 26; south. bank of Barkul' lake, on slope, No. 4932, Sept. 27; —1957, Kuan; south. slope of East. Tien Shan in Turfan basin, 10 km north of Kalangou settlement, nor. slope, 2700 m, No. 5807, June 23, 1958—Wang and Chang), *Jung. Gobi* (Baityk-Bogdo moutain range, nor.

slope, larch grove, Sept. 17, 1948—Grub.; Beidashan' [Baityk-Bogdo], No. 2425; on Tsitai-Beidashan' road, slope under shade, 1300 m, No. 5200—Sept. 28, 1957, Kuan).

General distribution: Jung.-Tarb., Arct., Europe (nor.-east. Europ. USSR), West. and East. Siberia, Nor. Mongolia, China (Altay).

L. dahurica Turcz. in Bull. Soc. natur. Moscou (1838) 101; Forbes and Hemsley, Index Fl. Sin. 2 (1902) 558; Mast. in J. Linn. Soc. London (Bot.) 37 (1906) 424; Wilson in Sargent, Pl. Wilson, 2 (1914) 21; id. Conif. and Taxads Japan (1916) 31; Komarov in Fl. SSSR, 1 (1934) 156. —*Abies gmelini* Rupr. Fl. Samojed. cisural. (1845) 56. —**Ic.:** Fl. SSSR, 1, Plate VIII, figs. 10–11.

Described from East. Siberia (Transbaikal). Type in Leningrad.

Forms sparse forests in different types of soils.

Found in border regions of Nor. Mongolia (east. of Mong.-Daur.) and in forest belt of B. Hinggan (around Arshan town).

General distribution: Arct., East. Sib., Far East, Nor. Mongolia (Mong.-Daur.), China (Dunbei), Korean peninsula.

3. **Pinus** L.
Sp. pl. (1753) 1000.

1. Cones narrowly ovate; umbo of ovuliferous scale without spinules but with recurved tubercle. Needles 5–7 cm long .... 1. **P. silvestris** L.
+ Cones broadly ovate, umbo of ovuliferous scale with slender acute spinules on tubercle. Needles 7–15 cm long ....
....2. **P. tabulaeformis** Carr.

1. **P. silvestris** L. Sp. pl. (1753) 1000; Ledeb. Fl. Ross. 3 (1851) 674; Krylov, Fl. Zap. Sib. 1 (1927) 80; Komarov in Fl. SSSR, 1 (1934) 167; Kitag. Lin. Fl. Mansh. (1939) 48; Grubov, Konsp. fl. MNR (1955) 57; Fl. Kazakhst. 1 (1956) 69; Dallimore and Jackson, Handb. Conif. (1964) 489. —*P. silvestris* var. *mongolica* Litv. in Schedae ad Herb. Fl. Ross. 5 (1905) 160; Danguy in Bull. Mus. nat. hist. natur. 20 (1914) 149; Gordeev and Jernakov in Acta pedol. sin. 2 (1954) 276; Wu in Acta phytotax. sin. 5, 3 (1956) 156. —**Ic.:** Fl. SSSR, 1, Plate VII, figs. 3–4, Plate VIII, fig. 8; Dallimore and Jackson, l.c. fig. 96.

Described from Scandinavia. Type in London (Linn.).

Rocky mountain slopes, sand masses along river valleys and in water divides.

IA. Mongolia: *Cis-Hing.* (forest strip north of Khalkha, sparce forest, 1899—Pal.; mountains east of Numurgin-Gol river [Khalkhin-Gol tributary], June 27, 1925—Kazakevich), *East. Mong.* ("Environs de Kailar, sables, alt. 800 m, No. 1364, June 22, 1894, Chaffanjon"—Danguy, l.c.; "left bank of Argun' river, in Khailar region, pine grove on sand"—Gordeev and Jernakov, l.c.).

General distribution: Aral-Casp., Fore Balkh.; Europe, Balk., Caucasus, West. and East. Sib., Far East, Nor. Mong. (Hent., Hang., Mong.-Daur.), China (Dunbei).

2. **P. tabulaeformis** Carr. in Traite Conif. ed. 2, 1 (1867) 510; Rehder and Wilson in J. Arn. Arb. 9 (1928) 7; Cheng in Contribs Biol. Lab. Sci. Soc.

China, Bot. ser. 6, 2 (1930) 14; Kitag. Lin. Fl. Mansh. (1939) 48; Walker in Contribs U.S. Nat. Herb. 28 (1941) 594; Ching in Bull. Fan mem. Inst. Biol. (Bot.) 10, 5 (1941) 260; Wu in Acta phytotax. sin. 5, 3 (1956) 155; Dallimore and Jackson, Handb. Conif. (1964) 497. —*Picea leucosperma* Maxim. in Bull. Ac. Sci. St.-Pétersb. 16 (1881) 558. —Ic.: Dallimore and Jackson, l.c. fig. 97.

Described from China. Type in Paris (?).

Forms montane forests, pure stands as well as admixed with broad-leaved aspecies.

IA. **Mongolia:** *Alash. Gobi* (Alashan mountain range: Tsuburgin-Gol gorge, April 28, 1908; Dyn'yuanin [Bayan-Khoto] oasis, in garden, March 20; same site, April 14 —1909, Czet.; Baisy monastery, spruce forest, 2800 m, July 6, 1957—Petr.; "Chien-Kou, Wang-Jeh-Fu, Ho-Lan-Schan, above 2600 m, l.c. Ching"—Walker, l.c.

IIIA. **Qinghai:** *Nanshan* (mountains on Tetung river, Sept. 3, 1872; same site, 2200 m, Aug. 6, 1880—Przew.; "Chui-Mo-Kou, Ching"—Walker, l.c.).

General distribution: China (Dunbei, North, North-west, Cent., East, South-west, South), Korean Peninsula (?).

Family 9. **CUPRESSACEAE** Bartl.

1. Shoots flattened, dorsiventral, branched in single plane. Mature cones woody, with divergent corymbiform scales, with uncinate external spinule .... 1. **Thuja** L. (*Th. orientalis* L.).
+   Shoots orbicular or tetrahedral, freely branched. Mature cones succulent, baccate, with connate fleshy scales .... 2. **Juniperus** L.

1. **Thuja** L.
Sp. pl. (1753) 1002.

1. **Th. orientalis** L. Sp. pl. (1753) 1002; Forbes and Hemsley, Index Fl. Sin. 2 (1902) 540; Walker in Contribs U.S. Nat. Herb. 28 (1941) 595. —*Biota orientalis* Endl. Syn. Conif. (1847) 47; Komarov in Acta Horti Petrop. 34, 1 (1920) 20; ibid in Fl. SSSR, 1 (1934) 192; Kitag. Lin. Fl. Mansh. (1939) 48; Fl. Tadzh. 1 (1957) 47. —Ic.: Fl. SSSR, 1, Plate 7, fig. 17; Der. i kust. SSSR [Trees and Shrubs of USSR], 1, fig. 78, 1–3.

Described from China. Type in London (Linn.).

In gorges, on rocks and slopes along with coniferous (pine) species and mixed forests in forest belt of mountains.

IA. **Mongolia:** *East. Mong.* ("Muni-Ula mountain range, south. slope, upper Ara-myrgyn gol, 1600 m, in mixed forest, July 5–12, 1871, Przew."—Komarov, l.c. [1920]; "Wula-Shan, Chien Kou, No. 3, 1940, Ching"—Walker, l.c.).

General distribuiton: China (Dunbei, North, North-west, Cent., East, South), Korean peninsula.

Note. Widely distributed ornamental tree grown even in inhabited regions of Central Asia: herbarium specimens of cultivated plants are available from Suidun and Kul'dzha (Junggar), Dynyuan'ina ([Bayan-Khoto] Mongolia) and Guidui (Qinghai).

2. Juniperus L.
Sp. pl. (1753) 1038; Gen. pl. (1737) 311.

1. Leaves in whorls of 3 each, acicular, identical. Cones subsessile, with 3 free seeds (subgenus *Oxycedrus* Spach) .... 2.
+ Leaves of adult plants scale-like, opposite in pairs, more rarely in whorls of 3 each; acicular in younger plants, in whorls of 3 each. Cones sessile at ends of oblong branches with 1–12 free seeds (subgenus *Sabina* Spach) .... 3.
2. Trees 1–8 m tall. Leaves long (2–2.8 cm long), narrowly linear, fine, rigid-cuspidate, erect, with sharply manifest keel beneath giving appearance of triquetrous leaves .... 1. **J. rigida** Sieb. et Zucc.
+ Robust shrubs with prostrate or even erect branches (very rarely small trees not taller than 1 m). Leaves short, 4–8 mm long, curved and appressed to their branches, short and acuminate, with obtuse keel beneath, grooved above with broad bright white central band .... 2. **J. sibirica** Burgsd.
3. Cones single-seeded .... 4.
+ Cones many-seeded .... 7.
4. Seeds rather oblong or ovoid, smooth or with furrows but without sculptured surface .... 5.
+ Seeds globose, with distinct sculptured surface .... 6.
5. Creeping shrubs with decumbent or ascending branches. Branches relatively slender, not more than 1.6 mm thick, usually 1.2–1.3 mm. Leaves rhombic or narrowly deltoid; their tips not or barely reaching upper leaves, with distinct dark-coloured oblong gland on back .... 6. **J. pseudosabina** Fisch. et Mey.
+ Trees up to 12 m tall or 2 m tall shrubs. Branches (1.6–2 mm) thick. Leaves broadly deltoid; their tips quite distinctly reaching upper leaves, without visible glands on back (covered by lower leaves) . ... 10. **J. turkestanica** Kom.
6. Leaves oblong, more or less long, acuminate, loosely adherent to branches, with orbicular light-coloured gland in lower 1/3 of leaf and often covered by lower leaf .... 5. **J. przewalskii** Kom.
+ Leaves rhombic, much smaller than in preceding species, subobtuse, very closely adherent to branches, with oblong dark-coloured gland in upper part of leaf and not covered by lower leaf .... 9. **J. tibetica** Kom.
7. Trees up to 10–12, sometimes even 18 m (more rarely 1–2 m tall shrubs) .... 8.
+ Creeping shrubs .... 9.
8. Branches lax, long, pendent, with predominant apical growth. Leaves rhombic, more or less obtuse or subobtuse, closely adherent to branches, with distinctly visible oval gland on back. Upper

part of seeds diverging at acute angle; as a result, cones truncated on top.. 8. **J. semiglobosa** Regel.

+   Branches short, not pendent. Leaves rather oblong, slightly acuminate, not closely adherent to branches, with indistinct oblong gland on back. Seeds closely appressed to each other throughout length; cones orbicular on top.. 4. **J. chinensis** L.

9.   Bark of branches reddish-grey. Branches compact, with predominance of lateral growth. Leaves elongate-rhombic, convex, with subobtuse or acuminate tip and distinctly visible oval gland on back, more or less appressed to branches. Seeds 1, 3, 4 or 6 ....7. **J. sabina** L.

+   Bark of branches ash-grey. Leaves ovate or ovate-lanceolate, acute or shortly acuminate and subacicular, with ovate or linearly lanceolate gland on back, more or less appressed to stem. Seeds 3–5, usually 3 .... 3. **J. arenaria** (Wils.) Florin.

Subgenus O x y c e d r u s Spach

1. **J. rigida** Sieb. et Zucc. Fl. Japon. 2 (1842) 57, non Nois. ex Desf. 1829, nom. nudum; Franch. Pl. David. 1 (1884) 292; Forbes and Hemsley, Index Fl. Sin. 2 (1902) 543; Wilson, Conif. and Taxads Japan (1916) 82; Komarov in Fl. SSSR, 1 (1934) 182; Kitag. Lin. Fl. Mansh. (1939) 49; Walker in Contribs U.S. Nat. Herb. 28 (1941) 594; Norlindh. Fl. mong. steppe, 1 (1949) 38; Lyu Shen'-o et al. Ill. fl. derev'ev i kust. S.-V. Kitaya [Illustrated Flora of Trees and Shrubs of North-East China] (1955) 102; Dallimore and Jackson, Handb. Conif. (1964) 270. —*J. utilis* Koidz. in Bot. Mag. (Tokyo) 44 (1930) 99; Iwata et Kusaka, Conif. Japon. (1952) 189. —**Ic.:** Sieb. et Zucc. l.c. tab. 127; Wilson, l.c. pl. LVIII; Iwata et Kusaka, l.c. tab. 71; Dallimore and Jackson, l.c. fig. 54.

Described from Japan. Type in Munich (?). Map 2.

Exposed mountain slopes, often as scrub in undergrowth of montane pine forests. Widely grown in countries of temperate latitudes.

IA. Mongolia: *East. Mong.* (Tres abondant dans toutes les montagne de l'Ourato, No. 2645, June 1866—David; environs mountain Shilin-Khoto town, hill steppe, 1959—Ivan.; Syrun-Bulyk mountain range [July 26–27, 1871]—Przew.; "Khongor-obo, stony mountain slopes, July 10, 1920, Eriksson, No. 718; eod. loco, Eriksson, Aug. 27, 1920, Nos. 337 and 869; eod. loco, in distr. montoso, June 9, 1931, Eriksson, No. 634"—Norlindh, l.c.), *Alash-Gobi* (Alashan range, in sparse forest, environs mountains Dyn'yuanin town [Bayan-Khoto], June 18, 1884—Przew.; same site, in gardens on clayey-sandy soil, March 16, 1908; same site, in garden, April 14, 1909; Tsuburgan-Gol gorge, April 28, 1908—Czet.; 30 km east of Bayan-Khoto town, Baisy monastery, juniper scrub, July 6, 1957—Petr.; same site, Fuiitin area, in undergrowth of pine forest, July 6, 1957—Kabanov; Wang Yeh Fu, alt. 1750–2125 m, May 4–16, 1923, Nos. 39 and 43—Ching; "Chien K'ou, No. 4; Shui Mo Kou, Ho Lan Shan, Nos. 93, 101, 1923, Ching"—Walker, l.c.).

General distribution: China (Dunbei, North), Korean peninsula, Japan (Khondo, Kyusyu).

2. **J. sibirica** Burgsd. Anleit. 2, Aufl. 2 (1790) 127; Komarov in Fl. SSSR, 1 (1934) 181; Kitag. Lin. Fl. Mansh. (1939) 49; Iwata et Kusaka, Conif. Japon. (1952) 192; Fl. Kirgiz. 1 (1952) 60; Grubov, Konsp. fl. MNR (1955) 57; Lyu Shen'-o et al. Ill. fl. derev'ev i kust. S.-V. Kitaya [Illustrated Flora of Trees and Shrubs of North-East China] (1955) 105; Fl. Kazakhst. 1 (1956) 72; Fl. Tadzh. 1 (1957) 49. —*J. nana* Willd. Berl. Baumz. (1796) 159; Turcz. Fl. baic.-dahur. 2 (1857) 145. —*J. communis* var. *nana* (Willd.) Loud. Arb. Brit. 4 (1838) 2486; Krylov, Fl. Zap. Sib. 1 (1927) 84. —*J. communis* auct. non L.: Simpson in J. Linn. Soc. London (Bot.) 41 (1912–1913) 444. —**Ic.**: Iwata et Kusaka, l.c. tab. 73; Fl. Tadzh. 1, Plate IV, fig. 3.

Described from Siberia. Type in Berlin.

In coniferous forests at upper limit of forest belt, in zone under bald peak, talus, placers, rocks, dry rocky slopes covered with shrubs; forms small thickets in Siberian sparse cedar and larch forests.

IA. Mongolia: *Mong. Alt.* (Upper Khobdos lake, upper Sagystei river, alpine meadow, June 30, 1906—Sap.; upper Kharagaitu-Gol, larch forest, No. 12993, Aug. 24, 1947—Yun.).

IIA. Junggar: *Cis-Alt.* (upward along Kandagatai river, Sept. 13, 1876—Pot.; larch forest on mountains on Khartsiktei bank, No. 69, July 27; mountain slope near Dzurkhe river, rare larch forest, No. 76, July 30—1898, Klem.; north of Burchum up to Kanas, 1400 m, No. 3015, July 18; Qinhe [Chingil'] in mountain gorge, 1900 m, Nos. 836 and 1210, Aug. 2; same site, Chzhunkhaitsza, forest, No. 1482, Aug. 7; Khasyungou, 1800 m, No. 2034, Aug. 18—1956, Ching; 18–20 km nor.-west of Shara-Sume, Kran basin, shrub, No. 1132, July 7, 1959—Yun.; same site, 1440 m, No. 10224, July 29; in Shintas region, sunny slope of mountain, 1800 m, No. 10764, July 20—1959, A.R. Lee (1959); "Larch forest zone from 1800–1950 m, Great Altai Mts. and Kran River, No. 133, Price"—Simpson, l.c.), *Tien Shan* (towards Koshety-Daban pass, near entrance to gorge, spruce forest fringe, May 23, 1877—Pot.; Sairam lake, July; Dzhagastai mountain, 1500–2100 m, Aug. 9; Talki river, 1800–2400 m, Aug. 16–1877, A. Reg.; Kokkamyr, 1800–2100 m, July; Kazan pass, Aug.—1878, A. Reg.; Kash river, Aryslan, 2700–3000 m, July 7, 1879—A. Reg.; Kara-Kary environs, nor. slope, spruce forest, June 23, 1893—Rob.; 2 km nor.-east of Sairam pass, on Urumchi-Kul'dzha road, fringes of spruce grove, Aug. 19, 1957—Yun.; 10 km south of mine in Dzhagastai, nor.-west. slope, No. 692, Aug. 8; nor. bank of Sairam lake, slope, 2800 m, No. 2135, Aug. 28; same site, southern bank, No. 1306, Sept. 1; Suidin-Utai, slope, No. 4137, Sept. 1—1957, Kuan), *Jung. Alt.* (Toli district, 2400 m, shaded side, Nos. 1229, 2662, 2985, Aug. 6, 1957—Kuan), *Tarb.* (region of Dachen [Chuguchak], shaded slope, 1770 m, No. 1627, Aug. 13—1957, Kuan).

General distribution: Jung.-Tarb., Tien Shan; Arct., Europe (north and hills of Cent. Europe), Asia Minor, Caucasus, Mid. Asia (Pamiro-Alay), West. and East. Sib., Far East, Nor. Mongolia (Fore Hubs., Hent., Hang.), China (Dunbei), Korean peninsula, Japan.

Note. The question of the distribution range of *J. sibirica* and its boundaries has not been solved thus far since much remains unknown about the differentiation of this species from *J. communis* and the correctness of its identification with other low-growing races (e.g. *J. nana* Willd.).

Subgenus S a b i n a Spach

3. **J. arenaria** (Wils.) Florin in Acta Horti Bergiani, 14 (1948) 353. —*J. chinensis* var. *arenaria* Wilson in Rehder and Wilson in J. Arn. Arb. 9 (1928) 20; Melle in Phytologia, 2 (1946) 195. —**Ic.**: Florin, l.c. Pl. 4, figs. 2–4.

Described from Tibet (Kuku-Nor lake). Type in Stockholm. In gorges, on dunes.

IIIA. Qinghai: *Nanshan* ("Kokonor Region: sand-dunes of Kokonor and along main lake on high dunes, alt. 3350 m, No. 13346, Sept. 1925, Rock"—Wilson, l.c.; "Richthofen Mountains, northern foothill at Hsiaohungku [Friis-Johansen, No. 2526, 1931]; north of Humboldt Range, mountain west of Tsaotaowannumen on river Tangho [Bohlin, No. 2258, 1932]"—Florin, l.c.).

General distribution: not reported from other sites.

Note. Specimens of this species were not available to us and hence it is difficult to comment on its genetic affinity and whether or not it deserves a specific rank. Melle (l.c.) studied all junipers included under the polymorphous species *J. chinensis* and concluded that *J. arenaria* was closer to *J. sabina* than to *J. chinensis*. This may be correct. For a more decisive view, material from Kuku-Nor and Nanshan is needed since the above-cited herbarium specimens are only available.

4. **J. chinensis** L. Mant. (1767) 127; Bunge in Mém. Ac. Sci. St.-Pétersb. Sav. Etrang. 2 (1835) 137; Franch. Pl. David. 1 (1884) 291; Forbes and Hemsley, Index Fl. Sin. 2 (1902) 539; Rehder et Wilson in Sargent, Pl. Wilson, 2 (1914) 60; Wilson Conif. and Taxads Japan (1916) 84; Rehder et Wilson in J. Arn. Arb. 9 (1928) 20; Hao in Engler's Bot. Jahrb. 68 (1938) 578; Ching in Bull. Fan mem. Inst. Biol. (Bot.) 10 (1941) 260; Walker in Contribs U.S. Nat. Herb. 28 (1941) 594; Melle in Phytologia, 2 (1946) 185; ej. Review *Juniperus chinensis* etc. (1947) 26; Florin in Acta Horti Bergiani, 14 (1948) 378; Lyu Shen'-o et al. Ill. fl. derev'ev i kust. S-V. Kitaya [Illustrated Flora of Trees and Shrubs of North-East China] (1955) 103; Dallimore and Jackson, Handb. Conif. (1964) 242.

Described from China. Type in London (Linn.). On rocks.

IA. Mongolia: *Alash.* Gobi (near Boro-Sondzhi collective, 135 versts [1 verst = 1.067 km—General Editor] north of Dyn'yuanin [Bayan-Khoto], July 20, 1873—Przew.; Dyn'yuanin oasis, in gardens, March 10; Alashan mountain range, Yamata gorge, May 1—1908, Czet.; "Ho Lan Shan, 1923"—Ching, l.c.; "Ha La Hu Kou, Nos. 52, 53; Pei Ssu Kou, No. 110, 1923, Ching"—Walker, l.c.).

IIIA. Qinghai: *Nanshan* ["Lien Cheng, often in rocky crevices on exposed cliffs, No. 320, 1923, Ching"—Walker, l.c.), *Amdo* (Guidetin [Guide] town, from sasa of Beisi temple, Oct. 8, 1908—Czet.).

General distribution: China (Dunbei—South, North, North-west, Cent., East, South-west, south), Korean peninsula, Japan.

Note. Highly polymorphous species. Many varieties and races have been described from China and Japan. Further, it is widely grown in China and Japan and hence it is difficult to establish the boundary of its natural occurrence.

5. **J. przewalskii** Kom. in Bot. mat. (Leningrad) 5 (1924) 28; Rehder and Wilson in J. Arn. Arb. 9 (1928) 18; Florin in Acta Horti Bergiani, 14, 8 (1948) 350; Dallimore and Jackson, Handb. Conif. (1964) 268. —*J. zaidamensis* Kom. l.c. 29; Rehder and Wilson, l.c. 18; Florin, l.c. 351. —*J. tibetica* auct. non Kom.: Rehder and Wilson, l.c. 19. —*J. glaucescens* auct. non Kom.: Rehder

and Wilson, l.c. 19. —*J. pseudosabina* auct. non Fisch. et Mey.; Walker in Contribs U.S. Nat. Herb. 28, 4 (1941) 594. —**Ic.**: Kom. l.c. 31, figs. 10, 11; Florin, l.c. Pl. 4, fig. 1.

Described from Tibet (Qinghai). Type in Leningrad. Plate III, fig. 1; map 2.

Forests along mountain gorges and isolated trees in alpine meadows, 3100–3900 m.

IC. Qaidam: mountains (Sarlyk-Ula mountains, Karagai-Gol river gorge, 3600 m, end of April, 1895—Rob.).

IIIA. Qinghai: *Nanshan* (mountains along Tetung river [Datunkhe], alpine zone, No. 318, July 1, 1872 [lectotypus!]; same site, Aug. 27, 1872; mountains along Tetung river, alpine meadow, May 6, 1873; Tetung river, July 20; Tetung river, 3000–3600 m, Aug. 1—1880, Przew.; Bara-Rdonsug river, 3100 m, May 6; Bardun river, Nagachir area, river valley, May 14; high mountains between Tashitu and Khsan rivers, May 25; Tashitu river, May 27, 1886—Pot.; Richthofen range and adjacent region: northern slopes of N. Kokonor barrier range, deep gorge or rocky slopes, alt. 3350 m, No. 13305, Oct. 1925—Rock; "Tai Hua, No. 558, forming pure forests or isolated on dry slopes, common, 1923, Ching"—Walker, l.c.; "Nanshan: Richthofen Moutnains, Hungshuipashango, alt. about 3700 m, Friis-Johansen, Nos. 2837 and 2871, 1931; northern foothill at Tahungchüan, about 65 km west of Suchow, Bohlin, No. 2259, 1932; Jüerhhung, foothill of mountain ridge east of river Suloho, Bohlin, No. 2257, 1932"—Florin, l.c.), *Amdo* (Kukunor mountain range, along Dulan-Gol river, 3300–3600 m, Feb. 23; upper Huang He, Dzhakhan-Fidza hill and Um river, April 19, 1880; South Kukunor mountain range, alongside Dulankhit pass, April 29, 1884—Przew.; same site, 3000–3600 m, May 3, 1895—Rob. Dulankhit temple, 3000–3900 m, Aug. 9, 1901—Lad.; grasslands between Labrang and yellow River; foot of sandstone cliffs in Serchen gorge; between Dzangar and Radja, No. 13918, May 15, 1925; Radja and Yellow River gorges; rocky slopes of Yellow River gorges between Nyavruch and Howa canyons, also Dacheo canyon, alt. 3340 m, No. 13946, May 27, 1926—Rock; "Radja and Yellow River gorges on mossy rocks in Howa valley, one stage north of Radja, alt. 3740 m, No. 14041, May 31, 1926; grasslands between Labrang and Yellow River; arid slope in Gochen valley near Yellow River gorge, south of Dzangar, alt. 3110 m, No. 13913, May 1926; Jupar range; Upper Jupar valley, rocky slopes, alt. 3700 m, No. 14302, June 27, 1926, Rock"—Florin, l.c.).

IIIB. Tibet: *Weitzan* {Amne-Machin mountain range, 4200 m, forests, Jan. 1895—Rob.; Burkhan-Budda moutain range, nor. slope, Khatu, Ikhe-Gol, Nomokhun gorges, 3300-3600 m, July 18, 1901—Lad.).

General distribution: China (North-west).

Note. A study of herbarium material and comparison of types convinced us that 4 species of the group of single-seeded junipers grow in Central Asia—*J. przewalskii* Kom., *J. pseudosabina* Fisch. et Mey., *J. tibetica* Kom. and *J. turkestanica* Kom.—although 6 species have been reported from the territory under study. *J. pseudosabina* and *J. turkestanica* are characterised by oblong or ovoid seeds, with glabrous or furrowed surface. *J. tibetica* and *J. przewalskii* differ from them in orbicular seeds with distinct, more or less profusely sculptured surface. *J. pseudosabina* is distributed in the west and north (Altay, Tien Shan, Junggar Gobi and Kashgar). Some reports of another single-seeded juniper, *J. turkestanica*, have been made from Kashgar (with its main distribution range in Pamiro-Alay). *J. tibetica* grows in Tibet (Weitzan) and North-west China while *J. przewalskii* grows in Qinghai and Nor. Tibet as well as in South-west China (see map 2).

V.L. Komarov has described one more species—*J. zaidamensis*—from Qaidam which, more correctly, could be considered a variety of *J. przewalskii*, to wit, *J. przewalskii* var. *zaidamensis*

(Kom.) Matz. comb. nova (Plate III, fig. 2). This variety differs from *J. przewalskii* in less acuminate leaves, very closely adherent to branches but with no distinct distribution range. Moreover, we detected in a herbarium specimen more or less acuminate small leaves, more or less adherent to the branches. Seeds of *J. przewalskii* var. *zaidamensis* can barely be differentiated from those of *J. przewalskii*.

Several species of the same group of single-seeded junipers have been described from the Chinese regions adjoining Cent. Asia by various investigators with varying degrees of reliability. Some perhaps do not merit the rank of species although a decisive opinion is difficult because the number of available herbarium specimens is small: in some cases, only type specimens are available while others, like the type specimens, are represented by 2–4 leaves. It must be pointed out that some species described from South-west and North-west China are highly similar to Central Asian species. Seeds and leaves of *J. glaucescens* Florin, *J. komarovii* Florin and *J. distans* Florin are very similar to those of *J. przewalskii* Kom. *J. potanini* Kom. has much in common with *J. pseudosabina* Fisch. et Mey. while *J. ramulosa* Florin is poorly distinguished from *J. tibetica* Kom.

6. **J. pseudosabina** Fisch. et Mey. in Index Sem. Horti Petrop. (1841) 65; Krylov, Fl. Zap. Sib. 1 (1927) 86; Komarov in Fl. SSSR, 1 (1934) 184; Persson in Bot. notiser (1938) 272; Fl. Kirgiz. 1 (1952) 60; Grubov, Konsp. fl. MNR (1955) 57; Fl. Kazakhst. 1 (1956) 72. —*J. turkestanica* Kom. in Bot. mat. (Leningrad) V (1924) 26, p.p. quoad pl. Asiae centr.

Described from Altay and Tarbagatai. Type in Leningrad. Plate II, fig. 1.

Alpine belt along rocky and rubble sites, talus, placers, bald peaks, scrub in subalpine belt, cedar and larch groves, as well as along upper forest limit.

IA. **Mongolia:** *Mong. Alt.* (mountain near bank of Kharchatyi [Khartsiktei], June 27, 1898—Klem.; Khan-Taishiri-Ula, nor. slope near creast, Sept. 21, 1945—Leont'ev; Bulugun river basin, Kharagaitu-Khutul' pass, alpine meadow and rubble placer with snow patches, July 24; upper Kharagaitu-Gol, left bank of Bulugun, larch forest, Aug. 24—1947, Yun.), *Gobi Alt.* (Ikhe-Bogdo mountain range, Bityuten-ama creek valley, alpine zone, Aug. 12, 1927—Sim. M.).

IB. **Kashgar:** *Nor.* (Talakyus canal, in Aksu, No. 8458, 2850 m, Sept. 23, 1958—Lee and Chu (A.R. Lee (1959)), *West.* (Jerzil, 3200 m, July 14, 1930, No. 110; Bostan-terek, about 3000 m, Aug. 11, 1934, No. 583"—Persson, l.c.; Sinkiang-Tibet highway, Ak-Kez daban pass, 3450 m, No. 535, June 5; along Kokshaal-Upal road, nor. slope, 2800 m, No. 226, June 10; along Sinkiang-Tibet highway, not far from Yarkend, in Shache region, 3110 m, No. 9894, July 15—1959, A.R. Lee (1959); Ak-Kez daban pass, south. slope, Raskem-Darya pass, rock crevices, July 5; King-Tau, 4 km south-east of Kosh-Kulak settlement, hill steppe along depressions on nor.-west. and nor. slopes; juniper groves with honeysuckle and dog-rose, July 10—1959, Yun. et al.).

IIA. **Junggar:** *Cis-Alt.* (Qinhe [Chingil'], 2250 m, No. 810, Aug. 3; in Koktogoi [Fuyun'] region, 3500 m, No. 1935, Aug. 7; same site, 2450 m, Aug. 17—1956, Ching; Timulbakhan area, sunny slope, 2000 m, No. 10741, July 18; in Shintas region, sunny slope, 1800 m, No. 10765, July 20—1959, A.R. Lee (1959)), *Jung. Alt.* (Toli district, 2400 m, No. 5786, Aug. 6; 17 km nor.-west of Ven'tsyuan' [Arasan], No. 1658, Aug. 29—1957, Kuan), *Tarb.* (Dachen [Chuguchak], shaded slope of hillock, Nos. 1623 and 1627, Aug. 13, 1957—Kuan), *Tien Shan.*

IIIC. **Pamir** (Ulug-Tuz gorge, right bank of Charlysh river, large scrub, June 27, 1909—Divn.).

General distribution: Jung.-Tarb., Tien Shan; Mid. Asia (West. Tien Shan and Trans-Alay mountain range), West. Sib. (East. Altay), East. Sib. (Sayans, Tannu-Ol, Transbaikal hills), Nor. Mongolia (Hent., Hang., Mong.-Daur.).

Note. *J. pseudosabina* is the most widely distributed single-seeded juniper of Cent. Asia, its distribution range the largest among the entire group of single-seeded junipers. *J. pseudosabina* forms scrub in Tien Shan, Tarbagatai, Altay, Sayans, Hangay and in mountains of Transbaikal. It has also been reported in East. Pamir and Kashgar. *J. turkestanica* Kom., closely related to *J. pseudosabina*, is found in our region only at some places in Kashgar (main distribution range of this species falls in Pamiro-Alay). Hybrids are frequent in the zone of contact of distribution ranges of these species (West. Tien Shan and Kashgar). Outwardly, they are very similar to *J. turkestanica* but differ in the presence of dark-coloured glands on leaves (on all or only some), rhombic leaves and other characteristics. Hybrids *J. pseudosabina* × *J. semiglobosa* are found in Kashgar.

Hooker [Fl. Brit. Ind. 5 (1890) 646] and Pampanini [Fl. Carac. (1930) 67] report the occurrence of *J. pseudosabina* in Karakorum, Kashmir, Sikkim and Bhutan. In the absence of herbarium specimens from these territories, this information is difficult to verify. At the same time, Hooker does not mention *J. pseudosabina* from our territory. The reference in the aforesaid works is perhaps to *J. wallichiana* Hook. f. et Thoms. ex Brandis found in Kashmir, Sikkim and Bhutan.

7. **J. sabina** L. Sp. pl. (1753) 1039; Simpson in J. Linn. Soc. London (Bot.) 41 (1912–1913) 444; Krylov, Fl. Zap. Sib. 1 (1927) 85; Komarov in Fl. SSSR, 1 (1934) 190; Fl. Kirgiz. 1 (1952) 67; Grubov, Konsp. fl. MNR (1955) 57; Fl. Kazakhst. 1 (1956) 74; Dallimore and Jackson, Handb. Conif. (1964) 271. —**Ic.:** Fl. Kazakhst. 1, Plate IV, fig. 4; Dallimore and Jackson, l.c. fig. 55.

Described from Italy. Type in London (Linn.).

Rocks and talus, granite outcrops, variously exposed rocky slopes, sand-dunes, valley floors and scrub steppes, spruce and larch groves, montane steppe and forest belts.

IA. Mongolia: *Mong. Alt.* (Barlyk [Barlagin-Gol] river gorge, April 23, 1877—Pot.; Bulugun river, July 25, 1898—Klem.; Onkattu lake, Chingistei post, alpine meadow, June 25, 1906—Sap.; 10 km nor. of Bidzhiin-Gol, foothill plain, scrub wasteland, July 17; 25–30 km south of Tamchi-Daba pass, midcourse of Bidzhiin-Gol river, left bank rocky slope, birch grove, Aug. 10—1947, Yun.; Bidzhiin-Gol river gorge near Tamchi-Daba pass, on rocks, Sept. 8, 1948—Grub.), *Cent. Khalkha* (Dzhargalante river, dunes, July 12, 1891—Levin; watershed between Ara-Dzhargalante and Ubur-Dzhargalante rivers, sand below Uste mountain, Sept. 17, 1925—Krasch. and Zam.; Ikhe-Tukhum-nor lake environs, Ongon-Khairkhan hill, June, 1926—Zam.; Sorgol-Khairkhan, along crevices in granite blocks, May 10, 1941—Yun.), *Val. Lakes* (hill on right bank of Bambukhei river—tributary of Tatsyn-Gol, July 21, 1893—Klem.), *Gobi-Alt.*, *Ordos* (Ulan-Morin river valley, sand-dunes, Aug. 21; dunes between Ushin camp and Kharukha-obo area, Sept. 3; dunes north of Narin-Gol, Sept. 10—1884, Pot.).

IB. Kashgar: *Nor.* (in Bai town region, mountain gorge, No. 8232, Sept. 7; nor.-west of Pochentszy, Mancher, 2100 m, No. 8333, Sept. 12—1958, Lee and Chu (A.R. Lee (1959)), *East.*

(Turfan district, 8 km north of San'shan'kou, forest steppe, 500 m, No. 5651, June 15; 10 km north of Kalangou, shaded slope, No. 5822, June 26—1958, Lee and Chu (A.R. Lee (1959)).

IIA. Junggar: *Cis-Alt.* (15 km from Qinhe [Chingil'], on arid slope, 1400 m, No. 923, 1360, Aug. 1; Altay district, 1200 m, No. 2482, Aug. 27—1956, Ching; 15 km nor.-west of Shara-Sume settlement, solonchak among granite lavas, scrub steppe, July 7; 18—20 km nor.-west of Shara-Sume, in Kran river basin, scrub meadow steppe, July 7—1959, Yun. et al.; nor.-west of Shara-Sume, Nos. 1440, 10223, April 29, 1959—A.R. Lee (1959); "Great Altai mts., Kran river and Upper Irtish mountain-sides from 1800 to 2900 m, No. 134, 1910, Price"—Simpson, l.c.), *Tarb.* (north of Dachen [Chuguchak] town, southern slope, No. 1516, Aug. 12; Dachen, shaded slope, 1770 m, No. 1701, Aug. 14—1957, Kuan), *Jung. Alt.* (Borguste river, 1800–3300 m, 1874—Larionov; Toli region along mountain peaks, 2150 m, No. 1080, Aug. 6; same site, Alabakzin mountain, slope, No. 2634, Aug. 7—1957, Kuan), *Tien Shan, Jung. Gobi, Zaisan* (Alkabek river, sand, Aug. 6, 1908—Fedtsch.).

IIIA. Qinghai: *Nanshan* (Kuku-Nor lake, 3060 m, on sandy soil, July 13, 1880—Przew.; dunes on southern bank of Dere-Nor lake [north of Kuku-Nor lake], 3000 m, April 25, 1886—Pot.; 70 km south of Chzhan'e town, Tsilin-Shan', spruce forest with undergrowth on nor. slope, 2600 m, July 12, 1958—Petr.).

General distribution: Fore Balkh., Jung.-Tarb., Cent. Tien Shan; Europe (South. and Cent. Europ. hills, south. Europ. USSR, Crimea), Caucasus, Mid. Asia, West. Sib. (Altay), Nor. Mongolia (Hent., Hang.), China (North, North-west).

8. **J. semiglobosa** Regel in Acta Horti Petrop. 6 (1880) 487; Komarov in Fl. SSSR, 1 (1934) 189; Fl. Kirgiz. 1 (1952) 63; Fl. Kazakhst. 1 (1956) 75; Fl. Tadzh. 1 (1957) 56. —*J. jarkendensis* Kom. in Bot. mat. (Leningrad) 4 (1923) 8. —? *J. chinensis* auct. non L.: Diels in Futterer, Durch Asien (1903) 7. —**Ic.:** Fl. Tadzh. 1, Plate III, fig. 2.

Described from Mid. Asia (Alay). Type in Leningrad.

Rocky, rubble and loessial slopes, rocks, along depressions, talus in middle and subalpine belts of mountains.

IB. Kashgar: *West.* (upper Yarkend-Darya, in Tokhta-khon area, 3300–3600 m, loess and rocks, Aug. 18, 1889—Rob.; King-Tau mountain range, 4 km south-east of Kosh-Kulak settlement, montane steppe, along depressions and talus on steep slope, June 10, 1959—Yun. et al.).

General distribution: Cent. Tien Shan; Mid. Asia (Pamiro-Alay and West. Tien Shan).

Note. An analysis of authentic material for *J. jarkendensis* Kom. and its comparison with authentic specimens of *J. semiglobosa* Regel convinced us of the total identity of these species. For this reason, *J. jarkendensis* Kom. has been placed among synonyms of *J. semiglobosa* Regel.

So far, *J. semiglobosa* has been reported only from Mid. Asia. This juniper species, new for Kashgar, is known to date only from 2 localities (upper course of Yarkend-Darya and King-Tau mountain range). Interestingly, the distribution of *J. semiglobosa* in Kashgar is contiguous with that of *J. turkestanica*. In Mid. Asia too these species occur together.

9. **J. tibetica** Kom. in Bot. mat. (Leningrad) 5 (1924) 27; Dallimore and Jackson, Handb. Conif. (1964) 279. —? *J. pseudosabina* auct. non Fisch. et Mey.; Strachey, Catal. (1906) 170. **Ic.:** Komarov, l.c. 31, fig. 9.

Described from Tibet. Type in Leningrad. Plate III, fig. 3; map 2.

Along gorges in forests.

IIIB. **Tibet**: *Weitzan* (Yangtze basin, both banks of river in Nruchu gorge, 3500 m, July 25; Mekong river basin, along Barchu river, 3600–4200 m, Sept. end—1900, Lad. [lectotypus!]).
General distribution: China (South-west).

10. **J. turkestanica** Kom. in Bot. mat. (Leningrad) 5 (1924) 26, quoad pl. e Pamiroalai et Tiansch. occid.: Komarov in Fl. SSSR, 1 (1934) 183, quoad pl. e Pamiroalai et Tiansch. occid.; Fl. Kirgiz. 1 (1952) 60, p.p.; Fl. Kazakhst. 1 (1956) 72, p.p.; Fl. Tadzh. 1 (1957) 51; Dallimore and Jackson, Handb. Conif. (1964) 279. —*J. centrasiatica* Kom. l.c. 27. —**Ic.**: Fl. Tadzh. 1, Plate IV, fig. 1.

Described from Mid. Asia (Turkestan mountain range). Type in Leningrad. Plate II, fig. 2.

Along rocky, rubble and melkozem slopes of mountains and in gorges; middle and upper forest belts, variously exposed slopes, forms compact stands, either pure or in admixture with *J. semiglobosa* Regel.

IB. **Kashgar**: *Nor.* (near Bedel' pass, 2700–3000 m, slopes and loessial descents of mountains, creeping shrub, June 10, 1889—Rob.), *West.* (nor. slope of Tokhta-khon mountain, 3600 m, July 30; same site, Sept. 2—1889, Rob.).
General distribution: Mid. Asia (Pamiro-Alay, West. Tien Shan).
Note. V.L. Komarov while describing *J. turkestanica* also describes in the same work another entire series of junipers including *J. centrasiatica* from Kashgar (Tokhta-khon area). A comparison of type specimens of *J. centrasiatica* and *J. turkestanica* confirms their similarity. Diagnosis of *J. centrasiatica* matches with that of *J. turkestanica*.
V.L. Komarov's diagnosis of *J. turkestanica* is not precise: it covers characteristics of *J. turkestanica* as well as of *J. pseudosabina*. Perhaps, V.L. Komarov erroneously placed a whole series of specimens of *J. pseudosabina* under *J. turkestanica* ("Pischpek, Kopal, Przewalsk, Kuldscha").

Class II. *GNETALES*
Family 10. **EPHEDRACEAE** Dum.

1. **Ephedra** L.
Sp. pl. (1753) 1040.

1.   Fruit scales[1] with broad scarious wings, dry when mature ....
    .... 2. **E. przewalskii** Sapf.
+   Fruit scales with only scarious border, sometimes almost invisible; succulent when mature .... 2.
2.   Fruits 2-seeded (single-seeded fruits extremely rare) .... 3.
+   All or most fruits single-seeded; occasionally 2-seeded fruits comprise about 1/2 of all fruits .... 7.
3.   Tubules of ovule integument straight or with single twist, up to 2 mm long (straightened) .... 4.

---

[1]Many investigators presently use different terminology for the reproductive organs of joint firs —strobilus, synangium, etc. We prefer the old terms. For brevity, we call the growing part of leaves together with teeth the sheath. It must be pointed out that morphologically differences in sheaths of the lower modes of branches among species of *Ephedra* are almost imperceptible and hence only upper nodes should be taken into consideration for species identification.

+ Tubules of ovule integument with many twists, often helical, 3–6 mm long (straightened) .... 6.

4. Cells in sheath coalescence zone and often in entire lower part with scarious cross-wrinkles (magnified) .... 11. **E. sinica** Stapf.

+ Cells in sheath coalescence zone whitish, more or less convex but without cross-wrinkles .... 5.

5. Fruits and male strobili (2) 3–5 each on common stalk or some arising singly in same plant; rarely, all 3–5 each on common stalk. Plant 10–20 (30) cm tall .... 1. **E. lomatolepis** Schrenk.

+ Fruits invariably single. Male strobili arising singly or occasionally 3 each on common stalk. Plant 3–7 (20) cm tall ....
.... 10. **E. regeliana** Florin.

6 (3). Branches glabrous .... 8. **E. glauca** Regel.

+ Branches scabrous .... 9. **E. intermedia** Schrenk et Mey.

7 (2). Seeds distinctly trigonous with scarious cross-wrinkles on surface. Shoots branching at 50–80°. Developed sheaths dish-shaped with very short obtuse teeth and broad rounded notch between them...
.... 6. **E. rhytidosperma** Pachom.

+ Seeds oblong-ovoid, biconvex or very faintly trigonous, glabrous. Shoots branching at 30–40°. Teeth of sheath acute, usually as long as connate part or slightly shorter, notch in zone of concrescence acute, deep; if teeth obtuse, notch in zone of concrescence more or less rounded, shallow, sheath cyathiform .... 8.

8. Fruits and male strobili singly in leaf axils. Plant 60–150 cm tall, with glabrous branches. Sheath reddish-brown or dark brown (old ones in zone of concrescence whitish), more or less thickened in lower part, fully developed with obtuse short teeth (not more than 1/4 sheath length). Inner pair of fruit scales connate for 2/3–3/4 length .... 3. **E. equisetina** Bunge.

+ Fruits and male strobili singly or more in leaf. Plant 3–20 (40) cm tall. Sheath yellowish (old ones whitish), semi-scarious, not thickened in lower part or slightly thickened under teeth, brownish; teeth acute, 1/2 sheath length or somewhat shorter. Inner pair of fruit scales connate for 1/4–1/2 (2/3) length .... 9.

9. Branches (1.2) 1.5–2 mm thick, sheath (3) 4 mm long, 2–3 mm broad. Tubules of ovule integument 0.5–1 mm long, straight ....
.... 7. **E. saxatilis** Royle ex Florin.

+ Branches 0.8–1 (1.2) mm thick, sheath 2–3 mm long, 1–1.5 mm broad. Tubules of ovule integument 1.5–2 mm long, straight or gyrose. ... 10.

10. Tubules of ovule integument gyrose. Branches finely scabrous, very rarely glabrous .... 5. **E. monosperma** J.G. Gmel. ex C.A. Mey.

+ Tubules of ovule integument straight. Branches glabrous ....
.... 4. **E. fedtschenkoae** Pauls.

Section A l a t a e Stapf.

1. **E. lomatolepis** Schrenk in Bull. phys.-math. Ac. Sci. St.-Pétersb. 3 (1844) 210; Stapf in Denkschr. Ak. Wien, math.-naturw. Kl. 56 (1889) 90; Bobrov in Fl. SSSR, 1 (1936) 197; Fl. Kazakhst. 1 (1956) 77, excl. pl. nonn. e reg. Balch.-Alak.; Pachomova in Opred. rast. Sr. Azii, 1 (1968) 28. —*E. stenosperma* Schrenk et Mey. in Mém. Ac. Sci. St.-Pétersb. 6 sér. 5 (1846) 267; Krylov, Fl. Zap. Sib. 1 (1927) 89. —*E. distachya* var. *stenosperma* Pavl. in Fl. Kazakhst. 1 (1956) 78 (in textu). —**Ic.:** Stapf, l.c. tab. 1, fig. 2; Fl. Kazakhst. 1, Plate 4, fig. 9.

Described from East. Kazakhstan (Lake Balkhash). Type in Leningrad. Sand, clayey-sandy and rubble soils of desert plains and foothills.

IIA. Junggar: *Jung. Gobi* (deserts between Gurban-Tangut and Sev-Kul'tei wells, Sept. 21, 1875—Pias.; between Qinhe and Fuyun' [Koktogoi], steppe plain, No. 752, July 31, 1956—Lee and Chu); south of Barbagai, 550 m, No. 2870, Sept. 8, 1956—Ching; east. bank of Ulyungur lake, embankment, sand, July 9; 15 km south of Ertai settlement along road to Guchen, baglur (*Anabasis brevifolia*) desert, July 16—1959, Yun.*; 15 km west of Kosh-Tologoi along road to Khobuk, area of low, rounded, isolated hills, June 22, 1957—Yun. et al.; Myaoergou, in steppe, No. 2430, Aug. 4, 1957—Kuan; 10 km north of Kosh-Tologoi along road to Altay, area of low, rounded, isolated hills, July 4, 1959—Yun.; Nor. Borborogusun, April 28, 1879—A. Reg.; bank of Kuitun river, No. 220, June 25, 1957—Shen-Tyan'; 15 km north of San'tszyaochzhuan [environs of Usu], on dunes, No. 1067, June 25, 1957—Kuan), *Zaisan* (left bank of Ch. Irtysh, Maikain area, hummocky sand, June 7, 1914—Schischk.), *Dzhark.* (Khorgos, April 22, 1877—A. Reg.; upper Ili, 750 m, sand bank [Kul'dzha region, July 15], 1877—Przew.; Ili valley between Chapchal and Dzhagastai, in wasteland along gorge, No. 3117, Aug. 7, 1957—Kaun*; 7–8 km south-west of Suidun settlement along road to Santokhodze quay, ridgy-hummocky sand, Aug. 31, 1957—Yun. et al.).

General distribution: Aral-Casp., Fore Balkh., Jung.-Tarb.; Mid. Asia (Kyzylkum, Syr Darya valley and Zeravshan, Gissar valley, spurs of Khozretish mountain range).

Note. Specimens of hybrid *E. lomatolepis* × *E. sinica* have been asterisked. See note under *E. sinica*.

2. **E. przewalskii** Stapf in Denkschr. Ak. Wien, math.-naturw. Kl. 56 (1889) 40; ? Paulsen in Hedin, S. Tibet, 6, 3 (1922) 98 quoad pl. e Bashkurgan; Florin in Norlindh, Fl. mong. steppe, 1 (1949) 38; Grubov, Konsp. fl. MNR (1955) 58; Chen and Chou, Rast. pokrov r. Sulekhe (1957) 86; Pachomova in Opred. rast. Sr. Azii, 1 (1968) 27. —*E. lomatolepis* var. *circinata* Regel in Acta Horti Petrop. 6 (1880) 484. —*E. kaschgarica* B. Fedtsch. et Bobr. in Bot. mat. (Leningrad) 13 (1950) 46. —*E. intermedia* auct. non Schrenk et Mey.: Pampanini, Fl. Carac. (1930) 68. —*E. intermedia* var. *glauca* auct. non Stapf; Walker in Contribs U.S. Nat. Herb. 28 (1941) 595; Florin in Acta Horti Berg. 14, 8 (1948) 355; Florin in Norlindh, Fl. mong. steppe, 1 (1949) 39. —*E. lomatolepis* auct. non Schrenk: Pavlov in Fl. Kazakhst. 1 (1956) 77, quoad pl. nonn. e reg. Balch.-Alak. —**Ic.:** Stapf, l.c. tab. 1, fig. 4.

Described from Mongolia. Type in Leningrad. Plate IV, fig. 3; map 3.

Pebble-sandy submontane plains, puffed solonchaks (rarely) and sand knolls; along open rocky, rubble and sand-covered slopes of knolls, flanks and floors of gullies and dry rocky gorges; up to 3000 m alt.

IA. Mongolia: *Mong.-Alt.* (Adzhi-Bogdo), *Bas. Lakes* (including northernmost report, nor. bank of Khirgis-Nur), *Val. Lakes, Gobi-Alt., East. Gobi, West. Gobi, Alash. Gobi* (including Khoir-tooroi village on Edzin-Gol river, July 18, 1886, Pot.—lectotypus!; extreme east. report: Bain-Ula mountains in Alashan range), *Khesi.*

IB. Kashgar: *Nor. West., South.* (including between Suget-bulak spring and Chizhgan-Darya river, May 28, 1885, Przew.—allolecto-typus!), *East., Takla-Makan* (Cherchen region), *Lob-Nor.*

IC. Qaidam: *plains* (Syrtyn, south.).

IIA. Junggar: *Jung. Gobi, Zaisan.*

General distribution: Aral-Casp., Fore Balkh., Jung. Alt.; Mid. Asia (Sultan-Uizdag, Fergana desert, Alay mountain range—Sokh river), Himalayas (Kashmir—Hunza).

Note. B.A. Fedtschenko and E.G. Bobrov assigned specimens of *E. przewalskii* with "glomerules and male and female strobili on common, 1–1.5 cm long stalks" to a separate species *E. kaschgarica* B. Fedtsch. et Bobr. We verified this feature in a very large amount of material from Cent. Asia as well as from Mid. Asia and concluded that the length of common stalk in *E. przewalskii* was variable throughout its distribution range (including Kashgar as well): from very short, visible only under a magnifying glass to stalks up to 1.5 cm in length. No correlation whatsover could be detected with other characteristics. It must also be pointed out that the common stalk in *E. strobilacea* Bunge is usually very short, occasionally reaching 1.5–3 cm in length. Thus there is no basis for treating *E. kaschgarica* B. Fedtsch. et Bobr. as a distinct species.

Hybrids *E. przewalskii × E. sinica* have been reported (Inner Mongolia, No. 29, May 4–6, 1923—Ching). This plant is similar to *E. przewalskii* in general appearance and type of furrows on branches but approaches *E. sinica* in branch rugosity. Transverse wrinkling of cells is barely visible in the zone of concrescence of sheaths. Pollen differs in size in a large quantity of undeveloped grains.

Section M o n o s p e r m a e Pachom.[2]

3. E. equisetina Bunge in Mém. Ac. Sci. St.-Pétersb. Sav. Etrang. 7 (1852) 500; Stapf in Denkschr. Ak. Wien, math.-naturw. Kl. 56 (1889) 81, saltem excl. pl. e Karagus.; Forbes and Hemsley, Index Fl. Sin. 2 (1902) 538; Danguy in Bull. Mus. nat hist. natur. 20 (1914) 148; Bobrov in Fl. SSSR, 1 (1934) 203, excl. pl. nonn. e Turkom.; Ching in Bull. Fan mem. Inst. Biol. (Bot.) 10, 5 (1941) 260; Walker in Contribs U.S. Nat. Herb. 28 (1941) 595; Florin in Norlindh, Fl. mong. steppe, 1 (1949) 39; Fl. Kirgiz. 1 (1952) 73; Grubov, Konsp. fl. MNR (1955) 57; Fl. Kazakhst. 1 (1956) 79; Pachomova in Opred. rast. Sr. Azii, 1 (1968) 32. —*E. procera* var. *typica* Regel in Acta Horti Petrop. 6 (1880) 480. —*E. procera* var. *typica* f. *cernua* Regel ex Kryl. Fl. Zap. Sib. 1 (1927) 91, saltem quoad pl. e Kaldschir. —*E. procera* auct. non Fisch. et Mey.:

---

[2]Galbuli plerumque monospermi. Semina ab utraque latere convexa vel trigonia. Vaginae urniformes vel crateriformes (rarius patelliformes). Antherae 7–8. Typus: *E. monosperma* J.G. Gmel. ex C.A. Mey.

Franch. Pl. David. 1 (1884) 284. —**Ic.:** Stapf, l.c. tab. 3, fig. 21, 3; Fl. Kazakhst. 1, Plate 4, fig. 10.

Described from Mid. Asia (Zeravshan mountain range). Type in Leningrad.

Stony and rubble desert and steppe slopes of mountains and knolls, talus and rocks, along stony flanks of gullies from foothill deserts to hilly steppes.

IA. Mongolia: *Mong. Alt., East. Mong.* (Datsin'shan'; nor. fringe of Kalgan), *Gobi-Alt., East. Gobi, Alash. Gobi, Ordos, Khesi* (east., Yunchan town vicinity).

IIA. Junggar: *Cis-Alt., Tien Shan, Jung. Gobi, Zaisan, Dzhark.*

General distribution: Jung.-Tarb.; Fore Asia (nor.), Caucasus, Mid. Asia, West. Sib. (Altay, south.), China (Dunbei, south., Nor.—nor. Shansi, Nor.-west.—nor. Shensi and Gansu).

4. **E. fedtschenkoae** Pauls. in Bot. Tidskr. 26 (1905) 254, excl. syn. *E. dubia*; Paulsen in Hedin, S. Tibet, 6, 3 (1922) 98; Florin in Kungl. Sv. Vet. Ak. Handl. 3 ser. 12, 1 (1933) 19; Bobrov in Fl. SSSR, 1 (1934) 202, excl. syn. *E. monosperma* var. *disperma* et tab.; Florin in Norlindh, Fl. mong. steppe, 1 (1949) 39, p.p.; Fl. Tadzh. 1 (1957) 83; Pachomova in Opred. rast. Sr. Azii, 1 (1968) 33. —*E. monosperma* auct. non Gmel. ex Mey.: Stapf in Denkschr. Ak. Wien, math.-naturw. Kl. 56 (1889) 73, p.p.; Forbes and Hemsley, Index Fl. Sin. 2 (1902) 539; Walker in Contribs U.S. Nat. Herb. 28, 4 (1941) 595. —*E. gerardiana* auct. non Wall. ex Stapf: Stapf in Denkschr. Ak. Wien, math.-naturw. Kl. 56 (1889) 75, p. min. p.; Florin in Kungl. Sv. Vet. Ak. Handl. 3 ser. 12, 1 (1933) 11, p. min. p. —**Ic.:** Paulsen, l.c. tab. 1, fig. 1.

Described from East. Pamir (Kara-Kul' lake). Type in Copenhagen.

Coastal pebble beds and rocks, rocky placers and rubble slopes, glacial moraines; middle and upper belt of hills.

IA. Mongolia: *Mong. Alt.* (island on Khobdos lake, on rocks, Aug. 2, 1899—Lad.; Tamchi-Daba pass, 2700 m, rocks, Sept. 7, 1948—Grub.).

IB. Kashgar: *West.* (10 km south of Torugart, 3400 m, No. 9755, June 20, 1958—Lee and Chu (A.R. Lee (1959)).

IIA. Junggar: *Tien Shan* (Nilki river, June 30, 1878; from Khatyn-Bogdo to Karashar, 2400 m, May 9; Irenkhabirga, lower Taldy, 1200 m, May 15; upper Taldy, 2400 m, May 17; same site, 2400 m, May 24; Borborogusun, 2400m, June—1879, A. Reg; Bedel' pass and uitas river gorge 2400–3800 m, upper watershed upland, May 31, 1889—Rob.; Khesho district, Bagrashkul' lake region, No. 7020, Aug. 3; same site, Bortu settlement, 2680 m, No. 7070, Aug. 4; Khetszin district, 4–5 km west of Bain-Bulak settlement, 2650 m, exposed slope, No. 6427, Aug. 10—1958, Lee and Chu (A.R. Lee (1959)); B. Yuldus, residual hill on right bank of Khaidyk-Gol, 4–5 km south-west of Bain-Bulak settlement, rocks, Aug. 10, 1958—Yun.).

IIIA. Qinghai: *Nanshan.* (in alpine belt of South Tetung mountain range, July 22–Aug. 5 1872—Przew.; along Bardun river, May 20, 1886—Pot.; La Che tzu Shan Mts, south of Sining, 3350–3900 m, No. 709, 1923—Ching), *Amdo* (in alpine belt of mountains along Mudzhik river, bank, June 28, 1880—Przew.; Radja and Yellow river, 3350 m, gorges among schist rocky slopes of river valley, No. 13939, May 27, 1926—Rock).

IIIB. Tibet: *Chang Tang.* (Sandzhu mountain range along road from Kheitszyan [on Raskem-Darya river] to Kirgiz-Dzhangal pass, near foothills of talus on bald slopes, June 2, 1959—Yun.; same site, 6 km east of Kheitszyan along highway, 4100 m, No. 465, June 2, 1954—Lee and Chu (A.R. Lee (1959)), *Weitzan* (Yantszytszyan river basin, Dychu river near Talachu estuary, 4000–4600 m, July 6, 1884—Przew.; south. slope of river basin watershed between Yantszytszyan and Mekong, 4000 m, Aug. 25, 1900—Lad.).

IIIC. Pamir (Ucha pass along clayey-stony descent, June 16, 1909—Divn.; "Mustagh-ata, Jam-bulak-bashi, 4439 m, Aug. 15, 1894, Hedin"—Paulsen, l.c.).

General distribution: Jung. Alt., Nor. and Cent. Tien Shan, East. Pam.; Mid. Asia (hilly region excluding Turkm. and much of Tadzh.).

5. **E. monosperma** J.G. Gmel. ex C.A. Mey. in Mém. Ac. Sci. St.-Pétersb. 6 ser. 5 (1846) 279; Stapf in Denkschr. Ak. Wien, math.-naturw. Kl. 56 (1889) 73, p.p.; Palibin in Tr. Troitskosavsko-Kyakht. otdelen. Priamursk. otd. RGO, 7, 3 (1904) 40; Danguy in Bull. Mus. nat. hist. natur. 20 (1914) 149; Rehder in J. Arn. Arb. 13 (1932) 385; Florin in Kungl. Sv. Vet. Ak. Handl. 3 ser. 12, 1 (1933) 118, p. max. p.; Bobrov in Fl. SSSR, 1 (1934) 202; Walker in Contribs U.S. Nat. Herb. 28 (1941) 595; Florin in Acta Horti Berg. 14, 8 (1948) 355; Florin in Norlindh, Fl. mong. steppe, 1 (1949) 39, p.p.; Grubov, Konsp. fl. MNR (1955) 58. —*E. minima* Hao in Contribs Inst. Bot. Nat. Ac. Peiping, 2, 7 (1934) 178; id. in Feddes Repert. 36 (1934) 197; id. in Engler's Bot. Jahrb. 68 (1938) 578. —**Ic.:** Mey. l.c. tab. 8, fig. 11; Stapf, l.c. tab. 3, figs. 19 and 31, 3; Fl. SSSR, 1 Plate 9, fig. 5.

Described from Siberia (Lena river). Type in Leningrad.

Pebble bed and pebble bed-sandy shoals, stony and rubble steppe slopes, rocks, from desert to alpine belt of mountains.

IA. Mongolia: *Mong. Alt.* ([Adzhi-Bogdo], Dzusylyk river, alpine belt, gravel, June 29; Dolonnor [Ker'-Nuru], granitic gravel, July 8—1877, Pot.; island rocks in Khobdo lake, Aug. 2, 1899—Lad.; Upper Khobdo lake, June 25, 1906—Sap.; Khan-Taishiri mountain range, Khabchigiin-Daba pass, 2275 m, Sept. 3, 1948—Grub.), *Cent. Khalkha* (nor. bank of Ugei-Nur near Khure, stones, June 15, 1893—Klem.; Boro-Undur mountain near Ugei-Nur lake, stony peak, July 16, 1924—Pavl.; Tsagan-Nur, Aug. 5, 1925—Glag.; Kholt area, June 6, 1926—Gus.; Khukhu-Khoshu area, Torgai river, July 24, 1926—Lis.), *Bas. Lakes* (Khirgis-Nur environs, Khudzhirtu south-east of Burgasytai river, dry pebble bed, July 27, 1879—Pot.), *Val. Lakes* (between Olon-Nur and Taishiri-Ula, in steppe, July 7, 1894—Klem.; around Dzasaktu-Khan camp [Yusun-Bulak], Aug. 8, 1930—Pob.), *Gobi Alt.* (Dzun-Saikhan, Aug. 24, 1931—Ik.-Gal.; same site, stony crest of mountain range, June 19, 1945—Yun.).

IIA. Junggar: *Tien Shan* (on Borborogusun river, 2700 m, June 15; Kunges, Arystan-Daban, 2700 m, Aug. 22—1879, A. Reg.), *Jung. Gobi* (valley of Khobuk river, deserted rubble slopes, July 20, 1914—Sap.; Myaoergou, slope, No. 2416, Aug. 4, 1957—Kuan).

IIIA. Qinghai: *Nanshan* (east. bank of Kuku-Nor lake near Ara-Gol river estuary, July 10, 1880—Przew.; along Bardun river, May 20, 1886—Pot.; Humboldt mountain range, Kuku-Usu river, Blagodatnyi cliff, 3600 m, July 7, 1894—Rob.; Kuku-nor like, sand bank, 3600 m, Aug. 29, 1908—Czet.), *Amdo* (Syansibei mountain range, May 28, 1880—Przew.).

IIIB. Tibet: *Weitzan* (Burkhan-Budda mountain range, nor. slope, Nomokhun-Gol gorge, 4250 m, May 26, 1900; Alyk-norin-Gol river valley, 3700 m, June 7, 1901—Lad.).

General distribution: Fore Balkh. (Kent mountains); West. Sib. (Altay), East. Sib., Far East (Ussur.), Nor. Mong.

Note. In regions in which *E. monosperma* and *E. fedtschenkoae* Pauls. grow together (Mong. Altay, Junggar), intermediate forms are found (combination of glabrous shoots with gyrose tubules or scabrous shoots with erect tubules). They are more like hybrids. Occasionally, however, such races are found within the distribution range of *E. monosperma* in which *E. fedtschenkoae* does not occur at present (e.g. in Transbaikal). It is quite likely therefore that these races reflect the intraspecific form evolutionary process. Field studies are needed to resolve this problem.

6. **E. rhytidosperma** Pachom. in Bot. mat. (Tashkent) 18 (1967) 51.

Described from Mongolia (Alashan mountain range). Type in Leningrad. Plate IV, fig. 1; map 3.

Rocky slopes of wastelands.

IA. **Mongolia:** *Alash. Gobi* (45 km south-west of Inchuan' town, rocky low mountains of Kholanshan' mountain range, montane semi-desert, July 5, 1957; 50 km south-west of Inchuan' town on Bayan-Khoto highway, south. part of Kholan'shan' mountain range, 1800 m, rocky slopes, June 10, 1958—Petr., typus!).

General distribution: endemic.

7. **E. saxatilis** Royle ex Florin in Kungl. Sv. Vet. Ak. Handl. 3 ser. 12, 1 (1933) 25. —*E. gerardiana* var. *saxatilis* Stapf in Denkschr. Ak. Wien, math.-naturw. Kl. 56 (1889) 76. —*E. gerardiana* auct. non Wall. ex Stapf: Hemsley in J. Linn. Soc. London (Bot.) 30 (1894) 118; ej. Fl. Tibet (1902) 198; Pampanini, Fl. Carac. (1930) 68, quoad pl. e Lanak-la.

Described from West. Himalayas. Type in Dehra Dun.

Granite rocks and talus, saline sand in lake regions, from 3000 to 5000 m.

IIIB. **Tibet:** *South.* (Khamba-la, 4850 m, 1904—Walton and Younghusband; in der Nahe von Lhassa, 3650 m, No. 23, 1921, R. Kennedy; Mt Everest Exp. 1922, yellow granite shelf SW of Rongbuk valley, 5000 m, June 22, 1922—E. Morton).

General distribution: Himalayas (Kashmir, west.).

Section E p h e d r a

8. **E. glauca** Regel in Acta Horti Petrop. 6 (1880) 480 and 484, quoad pl. e Nanschankou, p.p. (No. 1. ♂); Grubov, Konsp. fl. MNR (1955) 57; Pachomova in Opred. rast. Sr. Azii, 1 (1968) 31. —*E. intermedia* var. *glauca* Stapf in Denkschr. Ak. Wien, math.-naturw. Kl. 56 (1889) 62, excl. pl. turkestan.; Forbes and Hemsley, Index Fl. Sin. 2 (1902) 539. —*E. heterosperma* V. Nikit. in Fl. Tadzh. 1 (1957) 67, 503. —*E. intermedia* var. *tibetica* auct. non Stapf: Persson in Bot. notiser, 4 (1938) 273. —**Ic.:** Fl. Tadzh. 1, Plate 7, figs. 5–6.

Described from Kashgar. Type in Leningrad. Plate IV, fig. 2; map 4.

Pebble bed terraces and loessial basements on trails, debris cones, dry rocky gullies, rubble slope of moraines, rocky and stony slopes, from submontane to high-alpine (3750 m) wastelands.

IA. **Mongolia:** *West. Gobi* (Tsagan-Bogdo mountain range, upper belt, stony slopes, Aug. 4, 1943—Yun.; Shinshinsya environs, low rocky mountains near settlement, Oct. 5, 1959—Petr.), *Khesi* (south. fringe of Liyuan'in town along Lonsyr river, on rocks, July 4, 1886—Pot. [together with hybrids *E. glauca* × *E. sinica*]; Gaotai, in Gobi, No. 408, May 18; Tszyutsyuan'-Yuimyn', Nos. 414*, 415*, 422*, May 19, 1957—Kuan).

IB. **Kashgar:** *Nor.* (nor.-west. fringe of Bai depression, upper part of debris cone of Muzart river, on steep boulder slope, Sept. 6, 1958—Yun.; Bai depression, 2450 m, No. 8225, Sept. 7, 1958—Lee and Chu); between Kucha and Shakh'yar, 2300 m, No. 10073, July 27, 1959—Lee and Chu 15 km north of Shaikhly oasis [Aksu region], 2050 m, No. 8994x, Sept. 29, 1958—Lee and Chu), *West.* (including submontane desert between Upal and Bostan-Terek [settlements], stony bed, July 12, 1929, Pop.—allotypus! ♀), *South., East.* including Nanshankou, in rock crevices, June 1877, Pot.—typus, ♂).

IIA. Junggar: *Jung. Gobi* (30 km south-east of Urumchi, 1000–2400 m, Nos. 7372, 7379, June 21, 1958—Lee and Chu); Savan district, in San'daokhedza region, No. 350x, July 4; 10 km from Borotal [Bole], 720 m, No. 2172x, Aug. 29; Dashitou [Tsitai region]; No. 2099x, Sept. 24; nor. bank of Barkul' lake, on montane slope, No. 4418x, Sept. 26; north of Iu settlement, shaded slope, No. 2245x, Sept. 29—1957, Shen Tyan).

IIIA. Qinghai: Nanshan (Kuku-Usu river, July 10, 1879—Przew.; Humboldt mountain range, river valley, 2700–3050 m, arid clay, May 17, 1894—Rob. [together with hybrids *E. glauca* × *E. sinica*]; Altyntag mountain range, 15 km south of Aksai settlement, rocky slopes of gorge, 2800 m, Aug. 2, 1958—Petr.), *Amdo* (upper Huang He near Churmyn river estuary, May 23, 1880—Przew.*).

IIIC. Pamir (18 km south of Polur on road to Khoneiyailak, belt of pillow lavas, 3100 m, valley floor among boulders, May 12; upper course of Karakash river 10 km beyond Shakhidulla settlement along highway, 3750 m, high-alpine desert on debris cone, June 3; Gëz-Darya river valley, 90 km south of Upal oasis along road to Tashkurgan, June 11—1959, Yun.×).

General distribution: *Cent. Tien Shan* (Sarydzhas river basin), East. Pam. (West. Pshart river basin), Mid. Asia (Fergana, Shakhimardan river basin; Baisuntau mountain range Sangardak river basin).

Note. Hybrids *E. glauca* × *E. sinica* (asterisked) and *E. glauca* × *E. intermedia* (×-marked) are often found.

9. **E. intermedia** Schrenk et Mey. in Mém. Ac. Sci. St.-Pétersb. 6 sér. 5 (1846) 278; Bobrov in Fl. SSSR, 1 (1934) 198; Fl. Kirgiz. 1 (1952) 69; Fl. Kazakhst. 1 (1956) 77; Fl. Tadzh. 1 (1957) 74; Pachomova in Opred. rast. Sr. Azii, 1 (1968) 30. —*E. intermedia* var. *schrenkii* Stapf in Denkschr. Ak. Wien, math.-naturw. Kl. 56 (1889) 62. —*E. intermedia* var. *glauca* Stapf, ibid. 62, quoad pl. e Turkestan. —*E. intermedia* var. *tibetica* Stapf, ibid. 63, p.p.; Pampanini, Fl. Carac. (1930) 68. —*E. tibetica* (Stapf) V. Nikit. in Fl. Tadzh. 1 (1957) 70. —*E. microsperma* V. Nikit. in Fl. Tadzh. 1 (1957) 72, 503. —*E. ferganensis* V. Nikit. in Fl. Tadzh. 1 (1957) 73, 504. —*E. persica* V. Nikit. in Fl. Tadzh. 1 (1957) 73. —*E. valida* V. Nikit. in Fl. Tadzh. 1 (1957) 79, 504. —*E. glauca* auct. non Regel: V. Nikit. in Fl. Tadzh. 1 (1957) 68. —*E. przewalskii* auct. non Stapf: ? Paulsen in Hedin, S. Tibet, 6, 3 (1922) 98, quoad pl. e. Tsangpo. —**Ic.:** Stapf, l.c. tab. 2, fig. 15, 1; Fl. Tadzh. 1, Plate 7, figs. 1–4, 7, 8, Plate 8, figs. 1–3.

Described from East. Kazakhstan (Tarbagatai). Type in Leningrad.

Sand, sandy, rubble and stony slopes and rocks, from submontane wastelands to steppe belt of mountains.

IB. Kashgar: *West.* (Yarkend [without location or date], 1870, Henderson; possibly, specimen originated in Kashmir).

IIA. Junggar: *Cis-Alt.* (Bulgan somon, 14 km north-west of Ulyasutain-Gol estuary, on mountain top, 2700 m, July 25, 1947—Yun.), *Tien Shan* (from Sa-Daba to Urumchi, 1200 m, Nov. 3, 1879—A. Reg.; 18–20 km south of Urumchi along road to Davanchin, June 2, 1957—Yun.; 25 km south-east of Nyutsyuan'tsza, along bank of Nintszyakhe river, in forest, No. 85, July 17, 1957—Shen-Tyan; Dzinkho-Yakou, No. 4047, Sept. 1, 1957—Kuan), *Jung. Gobi* (Daofy area, May 23, 1879—Przew.; Sanchi [south-west of Urumchi], Nov. 6—1879—A. Reg.;

Baga-Khabtak-Nuru, nor. slope under main speak, stony slope, Sept. 14; same site, rocky slopes of dry gorge, 1500–1800 m, Sept. 14*; Tsetseg somon, 17 km west-north-west of Ubchugiin-Gol source, in area of low, isolated rounded hills on rocks, Sept. 9*; lower Khudzhirtiin-Gol, 20 km south-east of Oshigiin-Ulan-ula, along gully flanks, Sept. 16—1948, Grub.; Tushantszy-Karamai highway, No. 307, June 30; 5 km west of major bridge in Yan'tszykhai settlement, sunny slope, No. 450, July 7; south of Shichan, southern stony slope, No. 774, July 23; Fukan district, near east. fringe of Tyan'chi lake, No. 1893, Sept. 18; 36 km north of Barkul' lake, 2080 m, No. 2182, Sept. 26—1957, Shen Tyan), *Zaisan* (10 km west of Burchum along road to Zimunai, July 10, 1959—Yun.), *Dzhark.* (right bank of Ili west of Kul'dzha, April 1879—A. Reg.).

General distribution: Aral-Casp. (east., rare), Fore Balkh., Jung.-Tarb., Nor. and Cent. Tien Shan; Fore Asia (nor.), Mid. Asia, Himalayas (Kashmir).

Note. *E. intermedia* and *E. glauca* Regel, hybridise in regions of overlapping distribution. Hybrids are more often similar in general appearance to *E. glauca* but differ from it in faintly scabrous branches (see *E. glauca*). Hybrids *E. intermedia* × *E. sinica* (asterisked) too are found in Junggar.

10. **E. regeliana** Florin in Kungl. Sv. Vet. Ak. Handl. 3 ser. 12, 1 (1933) 17; Fl. Kazakhst. 1 (1956) 79; Fl. Tadzh. 1 (1957) 76; Ikonnikov in Dokl. AN Tadzh. SSR, 20 (1957) 55; Pachomova in Opred. rast. Sr. Azii, 1 (1968) 28. — *E. monosperma* var. *disperma* Regel in Acta Horti Petrop. 6 (1880) 479. —*E. pulvinaris* V. Nikit. in Fl. Tadzh. 1 (1957) 82, 504. —*E. monosperma* auct. non Gmel. ex Mey.: Stapf in Denkschr. Ak. Wien, math.-naturw. Kl. 56 (1889) 73, p.p.; Pampanini, Fl. Carac. (1930) 68, saltem quoad pl. e Braldo. —*E. gerardiana* auct. non Wall. ex Stapf: Stapf in Denkschr. Ak. Wien, math.-naturw. Kl. 56 (1889) 75, p.p. —*E. fedtschenkoae* auct. non Pauls.: Bobrov in Fl. SSSR, 1 (1934) 202, p.p.; Protopopov in Fl. Kirgiz. 1 (1952) 70 —*E. distachya* auct. non L.: Pavlov in Fl. Kazakhst. 1 (1956) 78, quoad pl. e reg. Balch.-Alak. —*E. minuta* auct. non Florin: V. Nikitin in Fl. Tadzh. 1 (1957) 78. —**Ic.:** Fl. Kazakhst. 1, Plate 4, fig. 8; Fl. Tadzh. 1, Plate 4, figs. 1–3.

Described from Nor. Tien Shan (Issyk-Kul' lake). Type in Leningrad.

Desert and desert-steppe rubble and rocky slopes, on rocks, along pebble bed and sandy banks of rivers, from plains to upper belt of mountains.

IB. Kashgar: *Nor.* (Muzart river valley 7-8 km beyond its emergence from hills into Baisk basin, 2000 m, desert-steppe belt, Sept. 7; same site, 2 km below Sazlik area, Sept. 9—1958, Yun.; in Aksu district, 2500 m, No. 8401, Sept. 12, 1958—Lee and Chu), *West.* (nor. slope of Tokhtakhon mountain, 2700–3050 m, along rocks, Aug. 18, 1889—Rob.; Kingtau mountain range near Koshkulak settlement, steppe belt, 2850 m, along washed slopes, June 10, 1959—Yun.), *East.* (Khetszin district, Ulastai region, 2500 m, No. 6337, Aug. 2; same site, west of Bain-Bulak, exposed slope, No. 6427, Aug. 10; same site, near river, No. 6508, Aug. 13—1958, Lee and Chu).

IIA. Junggar: *Tarb.* (along Tumanda river, Aug. 8, 1876—Pot.), *Tien Shan* (upper Borotala, 1800–2100 m, Aug. 14, 1878; Pilyuchi gorge, 900–1500 m, April 24, 1879; Nilki river, June 30, 1879—A. Reg.; Tekes river, 3600 m, dry steppe, June 3, 1893—Rob.), *Jung. Gobi* (30 km south of Shara-Sume on road to Shipati, desert steppe on loam, July 7, 1959—Yun.; Yan'ervo village near Urumchi, No. 554, May 21; lake in Urumchi region, in desert, No. 592, June 2—1957, Kuan; in der Umgebung von Urumtschi, Aug. 26–29, 1908—Merzb.), *Dzhark.* (Ili river west of Kul'dzha and around Kul'dzha, June 12, 1877; Taldy river gorge estuary, 900–1050 m, May 15, 1879—A. Reg.; Kul'dzha, 1878—Larionov).

IIIC. **Pamir** (Tashkurgan, on stony terrace, July 25, 1913—Knorr.; river valley 2–3 km south-west of Tashkurgan, along barren flanks of boulder end of debris, June 13, 1959—Yun.; same site, 3200 m, No. 314, June 13, 1959—Lee and Chu).

General distribution: Jung.-Tarb., Nor. and Cent. Tien Shan, East. Pamir; Mid. Asia (mountains, excluding Turkm. and south-west. Tadzh.), West. Sib. (Altay), Himalayas (Karakorum, west.).

Note. Transitional races to *E. sinica* Stapf are evidently of hybrid origin (see note under *E. sinica*).

11. **E. sinica** Stapf in Kew Bull. (1927) 133; Florin in Kungl. Sv. Vet. Ak. Handl. 3 ser. 12, 1 (1933) 4; Florin in Norlindh, Fl. mong. steppe, 1 (1949) 40; Grubov, Konsp. fl. MNR (1955) 58; idem in Bot. mat. (Leningrad) 19 (1959) 533. —*E. sinica* var. *pumila* Florin in Kungl. Sv. Vet. Ak. Handl. 3 ser. 12, 1 (1933) 11. —*E. vulgaris* auct. non Rich.: Hance in J. Bot. (London) 20 (1882) 295; Henderson and Hume in Lahore to Jarkand (1873) 336. —*E. distachya* auct. non L.: Kitag. Lin. Fl. Mansh. (1939) 49.

Described from China. Type in London (K).

Steppes and deserts, thin sand, stony and rubble slopes and rocks of mountains and knolls, along flanks and pebble beds of gullies, from submontane plains to midbelt of mountains.

IA. **Mongolia:** all regions except Khesi.

IIA. **Junggar:** *Jung. Gobi* (between Gurban-Tangut and Sevkyul'tei collectives, Sept. 21, 1875—Pias.; Uinchi, granites, Sept. 13; [lower] Bulugun river, Sept. 20—1930, Bar.; Manas river near Chendokhoze estuary, high terrace on right, May 28, 1954—Mois.; 5 km west of Yan'tszykhai, in desert, No. 307, June 30; 2–3 km south of Schichan, No. 830, July 24—1957, Shen Tyan; Tsitai-Meiyao, 700 m, in Gobi, No. 514, Sept. 25, 1957—Kuan*), *Zaisan* (between Burchum and Kaba river, Kiikpai well—Karoi area, sandy steppe, June 15, 1914—Schischk.).

IIIA. **Qinghai:** *Ambo* (Lonzhu river valley near Rtygri village, May 8, 1885—Pot.).

General distribution: West. Sib. (Altay), East. Sib. (south.), Nor. Mong. (Hang., Mong.-Daur.), China (Dunbei—west., Nor.—nor.-west., North-west—nor. Shensi, South-west—Sichuan, Daofu town).

Note. In the north-western part of the distribution range, *E. sinica* plants are more low-growing with more contorted shoots and shorter (due to growth constraint) internodes. Florin distinguished them as *E. sinica* var. *pumila* Florin, l.c. In some specimens from the same part of the distribution range, sheaths (in shape and barely visible transverse wrinkles in zone of concrescence) approach those of *E. regeliana* Florin: inner fruit scales are more concrescent. Some specimens stand so much midway between *E. sinica* and *E. regeliana* that it is very difficult sometimes to decide to which of these 2 species they belong. Such intermediate characteristics are particularly manifest in Altay where the distribution ranges of *E. sinica* and *E. regeliana* overlap. In the zone where these 2 species grow together, hybridisation processes evidently proceed leading to the formation of intermediate forms while the general appearance of *E. sinica* undergoes change in the north-western part of its distribution range (transitional specimens are asterisked). Field research is required to elucidate this aspect.

Without doubt, hybrids also exist between *E. sinica* and *E. przewalskii* Stapf (lone example, see under *E. przewalskii*), *E. sinica* and *E. glauca* Regel, *E. sinica* and *E. intermedia* Schrenk et Mey. Hybrids *E. sinica* × *E. glauca* are particularly numerous (West. and Alash Gobi, south. Kashgar—Keriya, Qaidam—Kuku-Nor lake region). These are characterised by a mixture of characteristics of parent species with all transitions between them. Hybrids differ from *E. sinica* in more vigorous growth, rather thick, poorly scabrous branches, gyrose tubules (predominance of characteristics of *E. glauca*), presence in same plant of seeds with erect and faintly gyrose ovules while bearing external similarity with *E. sinica* (predominance of characteristics of *E. sinica*); from *E. glauca* in poorly scabrous branches, transverse rugosity in concrescence zone of sheaths, frequently with very long teeth, etc. Hybrids *E. sincia* × *E. intermedia* have been detected in regions where parent species grow together (Gobi Altay, Junggar Gobi). The shape of their sheath represents a transition from *E. sinica* to *E. intermedia*; shoots, however, are contorted upward, which is not a feature of *E. intermedia*. All hybrids of *E. sinica* × *E. intermedia* examined were sterile (see *E. intermedia*).

Subdivision II. ANGIOSPERMAE

1. Embryo with single cotyledon. Vascular bundles of stem closed, without cambium and not thickened. Leaves mostly with parallel venation. Flowers usually comprise 5 trimerous verticels. .... Class I. **Monocotyledoneae**.
+ Embryo with 2 cotyledons. Vascular bundles of stem open, i.e., with cambial rings. Leaves with pinnate venation. Flowers usually with 5 pentamerous, more rarely dimerous verticels etc. .... Class II. **Dicotyledoneae**.

Class I. *MONOCOTYLEDONEAE*

1. Flowers without perianth or with vestigial perianth in form of glume (scale), hairs, bristles, glabrous or covered with bracts ..... 2.
+ Flowers with calyx and corolla or with simple perianth in form of corolla or calyx .... 12.
2. Flowers sessile in axils of scarious or scale-like bracts aggregated into spikelet; fruits invariably single-seeded. Inflorescence a spike, compound spike or panicle. Leaves linear, under part sheathing stem .... 3.
+ Flowers and inflorescence different .... 4.
3. Stem terete, in form of culm, with distinct nodes. Leaves distichous, margins of sheath tube not connate; distinct ligule (or rows of hairs instead) present at point of transition of sheath into blade, on its inner side. Fruit caryopsis with seed fused with fine membranous pericarp .... 19. **Gramineae** Juss.
+ Stem mostly trigonous, without nodes. Leaves tristichous, with closed sheaths and usually without ligule. Fruit paracarpous nutlet,

with seed not fused with pericarp; sometimes with pappus or surrounded by utricle .... 20. **Cyperaceae** Juss.

4. Flowers small, aggregated into compact cylindrical or globose spadix or flowers so sparse that plants appear flowerless .... 5.

+ Flowers do not form compact spadix .... 8.

5. Plants small, freely floating on water; comprise compact green leaf-like stems without leaves. Flowers appearing very rarely ....
.... 22. **Lemnaceae** S.F. Gray.

+ Plants with more or less developed stems and leaves .... 6.

6. Flowers small, densely aggregated on fleshy rachis into spadix. Stem grooved on one side, with acute rib on opposite side, transforming at site of spadix joint into wing, like leaf .... 21. **Araceae** Juss.

+ Rachis not fleshy, spadix cylindrical or orbicular .... 7.

7. Flowers aggregated into unisexual orbicular capitate inflorescences; upper staminate, lower pistillate. Fruit a nutlet with dry, spongy pericarp .... 12. **Sparganiaceae** Rudolphi.

+ Flowers aggregated into cylindrical spadix, with staminate flowers in upper and pistillate in lower part. Fruit a stalked nutlet with persistent style and stigma .... 11. **Typhaceae** Juss.

8 (4). Marsh or land plant. Perianth faintly visible, with scarious segments
.... 11.

+ Plants submerged in water. Perianth vestigial, sometimes substituted by enlarged appendages of stamens. Flowers single or aggregated into spicate inflorescences .... 9.

9. Small submerged herbs with opposite slender leaves. Flowers unisexual with 1 stamen or pistil .... 15. **Najadaceae** Juss.

+ Plants with leaves partly or fully submerged in water or partly floating, bearing membranous stipules at base .... 10.

10. Flowers bisexual or unisexual, monoecious, in spicate inflorescence with cylindrical rachis, not covered in upper leaf sheath; more rarely, single in leaf axil .... 13. **Potamogetonaceae** Dum.

+ Flowers aggregated into flat linear raceme enclosed during flowering in sheath of upper leaf .... 14. **Zosteraceae** Dum.

11 (8). Flowers, aggregated into terminal raceme. Leaves semi-cylindrical or flat, linear, glabrous. Fruit dry, schizocarp ....
.... 16. **Juncaginaceae** Rich.

+ Flowers aggregated into heads. Leaves subulate or flat, linear, with long hairs along margin. Fruit a trilocular capsule ....
.... 23. **Juncaceae** Juss.

12 (1). Flowers regular, actinomorphic .... 14.

+ Flowers irregular, zygomorphic .... 13.

13. Style united with filaments to form fleshy column, with anthers arranged symmetrically at its back (1 stamen generally developed

and united with style). Fruit a unilocular capsule, dehiscing by 6 oblong slits between placentas; seeds very small and numerous ... .28. **Orchidaceae** Juss.

+ Stamens free, only 3. Fruit a capsule, dehiscing by 3 valves and with diaphragms along valve centre .... 27. **Iridaceae** Juss.

14. Plants twining, dioecious. Flowers small, aggregated into axillary racemes; perianth with 6 segments, resembling sepals. Fruit a winged capsule or berry .... 25. **Dioscoreaceae** R. Br.

+ Plant with erect, not twining stems. Flowers usually bisexual, with more or less large corolliform perianth. Fruit a capsule without wings, aggregated ....15.

15. Ovary inferior ....16.

+ Ovary superior...17.

16. Stamens 6; plant bulbous ....26. **Amaryllidaceae** Jaume.

+ Stamens 3; plant with rhizome or corns.............. 27. **Iridaceae** Juss.

17. Carpels united; fruit a trilocular capsule or berry...24. **Liliaceae** Juss.

+ Carpels free, aggregated. Perennial aquatic or coastal grass ...... 18.

18. Inflorescence umbellate; stamens 9. Fruit with 6 many-seeded follicles, dehiscent along inner suture. Radical leaves sessile, trigonous ....18. **Butomaceae** Rich.

+ Inflorescence a pyramidal panicle; stamens 6. Carpels numerous, dry, single-seeded, not dehiscent, aggregated into pyramidal head. Radical leaves stalked .... 17. **Alismataceae** Vent.

Family 11. **TYPHACEAE** Juss.

1. **Typha** L.
Sp. pl. (1753) 973.

1. Staminate and pistillate inflorescences usually contiguous, without gaps between them. Leaves fairly broad, 10–20 mm broad .... 2. **T. latifolia** L.

+ Staminate and pistillate inflorescences usually separated from each other with more or less significant gaps on glabrous peduncle. Leaves narrow, 2–10 mm broad ....2.

2. Sheath at base of peduncle terminating in lanceolate cusp and without leaf blade .... 4. **T. minima** Funk.

+ Sheath at base of peduncle terminating in leaf blade ....3.

3. Pistillate inflorescence ovate or ovate-cylindrical, 4–8 cm long; pistillate flower without bracteoles; stigma spatulate .... 3. **T. laxmannii** Lepech.

+ Pistillate inflorescence cylindrical, 10–20 cm long; pistillate flower with bracteoles; stigma linear or lanceolate .... 1. **T. angustata** Bory et Chaub.

1. **T. angustata** Bory et Chaub. Exped. scientifiq. Moreè, 3, 2 (1832) 338; Kronf. Monogr. Typha (1889) 73; Graebn. in Engler, Pflanzenr. 2, IV, 8 (1900) 14; Forbes and Hemsley, Index Fl. Sin. 3 (1903) 171; Ostenf. in Hedin, S. Tibet, 6, 3 (1922) 97; B. Fedtsch. in Fl. SSSR, 1 (1934) 215; Kitag. Lin. Fl. Mansh. (1939) 50; Fl. Kirgiz. 1 (1952) 78; Fl. Kazakhst. 1 (1956) 32; Fl. Tadzh. 1 (1957) 86. —Ic.: Kronf. l.c. tab. IV, fig. 6; tab. V, fig. 1.

Described from Europe. Type in Florence (FI) (?).

Along coastal wet sites, in water.

IA. Mongolia: *East. Mong.* (on Huang He river, Aug. 8, 1884—Pot.), *Bas. Lakes* (environs of Khirgis-Nur lake, Aug. 10 and 14—1879, Pot.), *Ordos* (in Huang He river valley, July 20, 1871—Przew.; 75 km south of Dzhasak town, Aug. 17, 1957—Petr.).

IB. Kashgar: *West.* (Yarkend-Darya, 1050 m, June 15, 1889—Rob.; Kashgar oasis, Khan-Aryk vicinity, July 22, 1929—Pop.), *South.* (Niya, 1400 m, No. 9563, June 23, 1958—Lee and Chu), *Lob-Nor* ("Lob-Nor, Kara-Kochum, beneath Yust-chapghan, 816 m, June 25, 1900"—Hedin, l.c.).

IIA. Junggar: *Dzhark.* (Kul'dzha, June 15, 1877—A. Reg.; Suidun, July 16, 1877—A. Reg.; same site, March 1878—Aliakhun; same site [July 3, 1886]—Krasnov).

General distribution: Aral-Casp., Fore Balkh., Nor. Tien Shan; Europe (lower course of Volga), Mediterranean, Balk.-Asia Minor, Fore Asia, Mid. Asia, China (Dunbei, North, Northwest, ? South-west), Japan.

2. **T. latifolia** L. Sp. pl. (1753) 971; Kronf. Monogr. Typha (1889) 90; Graebn. in Engler, Pflanzenr. 2, IV, 8 (1900) 8; Forbes and Hemsley, Index Fl. Sin. 3 (1903) 172; Danguy in Bull. Mus. nat. hist. natur. 20 (1914) 142; Krylov, Fl. Zap. Sib. 1 (1927) 93; B. Fedtsch. in Fl. SSSR, 1 (1934) 210; Kitag. Lin. Fl. Mansh. (1939) 50; Fl. Kirgiz. 1 (1952) 77; Fl. Kazakhst. 1 (1956) 81; Fl. Tadzh. 1 (1957) 85. —Ic.: Kronf. l.c. tab. V, fig. 11; Pflanzenr. 2, IV, 8, fig. 3A; Fl. Kazakhst. 1, Plate V, fig. 2.

Described from Europe. Type in London (Linn.).

In shallow waters of rivers and lakes, along marshes, river backwaters and meanders.

IIA. Junggar: *Dzhark.* (environs of Suidun [July 13, 1886]—Krasnov), *Jung. Gobi* ("Bords de rivieres entre l'Ebi-Nor et la vallée de l'Irtich, March 17, 1895"—Danguy, l.c.), *Zaisan* (Chenkur [Burchum] river valley, Aug. 17, 1906—Sap.).

General distribution: Aral-Casp., Fore Balkh., Tien Shan; Europe, Mediterranean, Balk.-Asia Minor, Fore Asia, Caucasus, Mid. Asia, West. and East. Sib., Far East, China (Dunbei), Japan, North America.

3. **T. laxmannii** Lepech. in Nova Acta Ac. Petrop. 12 (1801) 84, 335; Franch. Pl. David. 1 (1884) 312; Kronf. Monogr. Typha (1889) 81; Graebn. in Engler, Pflanzenr. 2, IV, 8 (1900) 10; Forbes and Hemsley, Index Fl. Sin. 3 (1903) 172; Krylov, Fl. Zap. Sib. 1 (1927) 94; B. Fedtsch. in Fl. SSSR, 1 (1934) 212; Norlindh. Fl. mong. steppe, 1 (1949) 41; Fl. Kirgiz. 1 (1952) 77; Grubov, Konsp. fl. MNR (1955) 58; Fl. Kazakhst. 1 (1956) 81; Fl. Tadzh. 1 (1957) 85. —*T. stenophylla* Fisch. et Mey. in Bull. Ac. Sci. St.-Pétersb. 3 (1845) 209. —Ic.: Lepech. l.c. tab. IV; Kronf. l.c. tab. IV, fig. 3, tab. V, fig. 15.

Described from Transbaikal. Type in Leningrad. Plate V, fig. 1.

Wet and marshy areas, particularly common on banks of rivers, lakes and irrigation canals.

IA. Mongolia: *Cis-Hing.* (Khalkhin-Gol, Symbur area, Sept. 1, 1928—Tug.), *Cent. Khalkha* (Tsagan-Nur lake, in water along banks, June 28, 1924—Pavl.), *East. Mong.* (Khailar town, July 6, 1901—Lipsk.; Baishintin-Sume vicinity, marsh, Aug. 17, 1927—Zam.; "12 km ad orient. vers a Doyen, in laculo, 31 Jul. 1924, No. 40, Eriksson"—Norlindh, l.c.), *Val. Lakes* (Orok-Nur, wet lacustrine lagoons and solonetzes, Aug. 4, 1926—Tug.), *Alash. Gobi* (Bain-Nor lake, Sept. 2, 1884—Pot.; Dyn'yuan'in [Bayan-Khoto] oasis, on silty soil, June 22, 1908—Czet.; Dzin'ta district, Sadivan' village, low solonchak plain, July 18, 1958—Petr.), *Ordos* (10 km south-west of Ushin town, in lake among willow rushes, Aug. 4; 75 km south of Dzhasak town, Tautykhai lake, in water, Aug. 17—1957, Petr.), *Khesi* (Gan'chzhou, July 27, 1875—Pias.; Lonsyr river valley, June 2, 1886—Pot.).

IB. Kashgar: *Nor.* (Uchturfan, 1886—Krasnov), *West.* (Kashgar oasis, Yadoma village, along irrigation canals, July 23; environs of Yangishar, July 25—1929, Pop.), *Takla-Makan* (Cherchen-Darya, April 1885—Przew.), *Lob-Nor* (saline mire [near Machan-Ul] mountain, July 28, 1879—Przew.).

IC. Qaidam: *plains* (Erdeni-Obo area, silty floor of marsh overlain with reed remnants, 3000 m, Aug. 7, 1901—Lad.).

IIA. Junggar: *Tien Shan* (Kash river, 900 m, Sept. 6, 1878; same site, 1350 m, May 2, 1879—A. Reg.), *Jung. Gobi* (Bulun-Tokhoi and Urungu river, sand, Aug. 19, 1876—Pot.), *Zaisan* (sand on right bank of Ch. Irtysh, near Kaba estuary, Aug. 19, 1906—Sap.; Zimunai, 650 m, No. 10584, June 25, 1959—A.R. Lee).

General distribution: Aral-Casp., Fore Balkh., Jung.-Tarb., Tien Shan; Europe (south. Europ. USSR), Mediterranean, Balk.-Asia Minor, Caucasus, Mid. Asia, West. Sib. (south); East. Sib. (west), Far East (south), Nor. Mong. (Hang., Mong.-Daur.), China (Dunbei, Nor.), Japan.

4. **T. minima** Funk in Hoppe, Bot. Taschenb. (1794) 118, 181 (nomen); Schnizl. Typhac. (1845) 25; Franch. Pl. David. 1 (1884) 313; Kronf. Monogr. Typha (1889) 58; Graebn. in Engler, Pflanzenr. 2, IV, 8 (1900) 14; Forbes and Hemsley, Index Fl. Sin. 3 (1903) 172; Danguy in Bull. Mus. nat. hist. natur. 17 (1911) 450; B. Fedtsch. in Fl. SSSR, 1 (1934) 216; Kitag. Lin. Fl. Mansh. (1939) 50; Walker in Contribs U.S. Nat. Herb. 28 (1941) 595; Fl. Kirgiz. 1 (1952) 81; Fl. Kazakhst. 1 (1956) 84; Fl. Tadzh. 1 (1957) 88; Chen and Chou, Rast. pokrov r. Sulekhe (1957) 91. —*T. angustifolia* β *minor* L. Sp. pl. ed. 2 (1762) 1378. —*T. laxmannii* auct. non Lepech.: Ledeb. Fl. Ross. 4 (1853) 3. —*T. pallida* Pobed. in Bot. mat. (Leningrad) 11 (1949) 17; Fl. Kazakhst. 1 (1956) 84; Fl. Tadzh. 1 (1957) 88. —**Ic.:** Kronf. l.c. tab. IV, fig. 2, tab. V. fig. 7; Fl. Kazakhst. 1, Plate V, fig. 1.

Described from Europe. Type in Vienna.

Along banks of rivers and lakes, in swamps.

IA. Mongolia: *East. Mong.* (Khailar town vicinity, in water, No. 1140, July 4, 1951—S.H. Li et al. (1951)), *Alash. Gobi* (south. Gobi [Edzin-Gol river, May–June] 1886—Pot.; "Chung Wei, No. 229, on margins of streams, Ching"—Walker, l.c.), *Khesi* (on Edzin-Gol river, vicinity of Luyatun village, June 18; Gaotai, June 20—1886, Pot.; Sachzhou [Dun'khuan], June 25, 1879— Przew.; Sachzhou, moist meadow, 1700 m, March 5, 1891—Rob.; "Lin-Chouei, marai, alt. 1500 m, environs de Sou-tcheou, June 25, 1908, Vaillant"—Danguy, l.c.; Sulekhe river—Chen and Chou, l.c.).

IB. Kashgar: *West.* (Yarkend-Darya river, along streams, 900 m, June 22, 1889—Rob.), *South.* (Niya, in wet sites, June 4, 1885—Przew.; 120–125 km south of Khotan, 1140 m, No. 9630, May 18, 1959—A.R. Lee), *Takla-Makan* (Khotan, Machzha-shan', 1100 m, No. 9635, May 22; Cherchen district, in Tatrak region, in river floodplain, 1213 m, June 2—1959, A.R. Lee).

IIA. Junggar: *Tien Shan* (Talki area, July 18, 1877—A. Reg.; Tekes, wet clay, near river, July 10, 1893—Rob.), *Dzhark.* (Kul'dzha, May 3, June, July 6—1877, A. Reg.).

IIIA. Qinghai: *Amdo* (Churmyn river, 2700–2850 m, May 17; marshy sites on Huang He river, June 2—1880, Przew.).

General distribution: Fore Balkh., Europe (Cent. and West.), Mediterranean, Balk.-Asia Minor, Fore Asia, Caucasus, Mid. Asia, East. Sib. (Tuva), Nor. Mong. (?), China (Dunbei, North).

Note: While describing the new species *T. pallida*, closely related to *T. minima*, E.G. Pobedimova (l.c.) distinguished it from the latter as bearing orbicular female inflorescence or elliptical, numerous isabelline carpodia, often slightly larger than hairs of gynophore, bracteoles and stigma of fertile flowers, and hairs not enlarged at tip. Let us examine these features.

Analysis of herbarium specimens showed that the shape of the female inflorescence "suborbicular or elliptical" (*T. pallida*) and ovate or shortly cylindrical (*T. minima*) is an extremely unstable characteristic for differentiating the species. The shape of the inflorescence shows great variation even within the same plant. Moreover, shape, size, colour and surface structure of the inflorescence depend on conditions of plant collection. This should be borne in mind not only in the case of these species, but also for distinguishing all cattails. An in-depth study (extremely necessary) of this genus is possible only in the field. The next feature—number of carpodia and their length—is also unreliable for differentiating the species since it is subject to considerable fluctuations in these 2 "species". The next one, i.e., hairs of gynophore, would appear to be a distinct characteristic. According to the description of E.G. Pobedimova, gynophore hairs of *T. pallida* are not enlarged at the tip, in which respect it differs from *T. minima*. However, even in the type specimen of *T. pallida*, most hairs of gynophore are enlarged. The same feature is also visible in herbarium specimens labelled *T. pallida*. The foregoing discussion thus compels us to consider *T. pallida* Pobed. a synonym of *T. minima*.

Family 12. **SPARGANIACEAE** Rudolphi

1. **Sparganium** L.
Sp. pl. (1753) 971.

1. Inflorescence simple, not branched; bracts slender, colourless or pale; heads sessile. Carpels with long erect beak (considerably longer than carpel) .... 2. **S. simplex** Huds.

+ Inflorescence more or less branched; bracts, at least in central part, compact, dark-coloured; heads stalked, on lateral branches or on main rachis (*S. ramosum* Huds. s. ampl.) .... 2.

2. Carpels obpyramidal, sessile, abruptly transformed at tip into beak .... 3. **S. stoloniferum** Buch.-Ham.

+ Carpels fusiform, stalked, gradually transformed at tip into beak .... 1. **S. microcarpum** (Neum.) Čelak.

1. **S. microcarpum** (Neum.) Čelak. in Österr. bot. Zeitschr. 46 (1896) 423; Juzepczuk in Fl. SSSR, 1 (1934) 221; Fl. Kirgiz. 1 (1952) 82; Fl. Kazakhst. 1 (1956) 86. —*S. ramosum* f. *microcarpum* Neum. in Hartm. Scand. Fl. 12 Uppl. (1889) 112; Graebn. in Engler, Pflanzenr. 2, IV, 8 (1900) 15; Krylov, Fl. Zap. Sib. 1 (1927) 97. —**Ic.:** Čelak. l.c. tab. VIII; Fl. SSSR, 1 Plate XI, fig. 4a–c).

Described from Cent. Europe (Czechia). Type in Prague.

On water along banks of lakes, ponds, slow-moving brooks, marshy floodplain meadows, along swamps.

IIA. Junggar: *Dzhark.* (Kul'dzha, June 15, 1877—A. Reg.; near Suidun [July 1886]—Krasnov).

General distribution: Aral-Casp., Fore Balkh., Jung. Tarb., Tien Shan; Europe (except south.), Caucasus, Mid. Asia, West. Sib. (south).

Note. *S. microcarpum* and *S. stoloniferum* represent complex species of inadequately studied group *S. ramosum* Huds. sp. coll. (cycle Ramosae Juz.). Often, when it is not possible to accurately identify the species of this group, the collective name *S. ramosum* is used. Hence even the geographic distribuiton of the two species is rather vague.

2. **S. simplex** Huds. Fl. Angl. ed. 2 (1778) 401; Schnizl. Typhac. (1845) 26; Ledeb. Fl. Ross. 4 (1853) 4; Graebn. in Engler, Pflanzenr. 2, IV, 8 (1900)16; Danguy in Bull. Mus. nat. hist. natur. 20 (1914) 143; Krylov, Fl. Zap. Sib. 1 (1927) 99; Juzepczuk in Fl. SSSR, 1 (1934) 223; Kitag. Lin. Fl. Mansh. (1939) 51; Norlindh, Fl. mong. steppe, 1 (1949) 42; Fl. Kirgiz. 1 (1952) 82; Grubov, Konsp. fl. MNR (1955) 58; Fl. Kazakhst. 1 (1956) 86. —**Ic.:** Pflanzenr. 2, IV, 8, fig. 3k; Fl. SSSR, 1, Plate XI, fig. 8a–f.

Described from England. Type in London (BM) (?).

On coastal shallow waters of rivers, lakes, brooks, ponds, etc., occasionally in deep waters forming submerged races with floating leaves.

IA. Mongolia: *Mong. Alt.* (Dain-Gol lake, in water, July 29, 1909—Sap.), *East. Mong.* (near Chingiskhan railway station, bank of lake, July 23, 1903—Litw.; Khailar town, in water, No. 2972, 1954—Wang; "Doyen, in aqua stagnante, Aug. 5, 1926, No. 228, Eriksson"—Norlindh, l.c.), *Bas. Lakes* (environs of Khirgis-Nur lake, Aug. 18, 1879—Pot.; "stream between Khara-Nur and Khara-Usunur"—Grubov, l.c.).

General distribution: Aral-Casp., Fore Balkh.; Europe (except extreme south), Fore Asia, Caucasus, Mid. Asia (occasionally in northern part), West. Sib., East. Sib. (cent.), Far East, Nor. Mong. (Hang.), China (Dunbei), Korean peninsula (North), India (north.), North America.

Note. Highly variable species. Races differing considerably from the type specimen are often found. The aquatic form of this species with floating leaves and, frequently, a shorter rachis of male inflorescence is known as *S. longissima* Fritsch.

3. **S. stoloniferum** Buch.-Ham. in Wallich, Cat. (1832) No. 4990, nomen; Graebn. in Engler, Pflanzenr. 2, IV, 8 (1900) 14, descr.; Juzepczuk in Fl. SSSR, 1, (1934) 219; Kitag. Lin. Fl. Mansh. (1939) 51; Grubov, Konsp. fl. MNR (1955) 58; Fl. Kazakhst. 1 (1956) 85; Fl. Tadzh. 1 (1957) 90. —**Ic.:** Pflanzenr. 2, IV, 8 fig. 3c; Fl. SSSR, 1, Plate XI, fig. 2a–b.

Described from East. India. Type in Calcutta.

IA. **Mongolia:** *Ordos* (Huang He river, Aug. 6, 1871—Przew.; Narin-Gol river, Sept. 11, 1884—Pot.).

General distribution. Aral-Casp., Fore Balkh.; Fore Asia, Mid. Asia, West. Sib. (south), East. Sib. (Daur.), Far East (south.), Nor. Mongolia (Mong.-Daur.), India, ? Japan.

Family 13. **POTAMOGETONACEAE** Dum.

1. Flowers singly or in pairs in leaf axils, unisexual, monoecious; at least female flowers enveloped in scarious bract ....
   .... 3. **Zannichellia** L.
+ Flowers in spicate inflorescence (sometimes only with 2 flowers), bisexual, without perianth .... 2.
2. Inflorescence a many-flowered 'spike', cylindrical or orbicular; stamens 4, with large connective appendage resembling perianth segments. Carpels sessile .... 1. **Potamogeton** L.
+ Inflorescence a 2-flowered 'spike'; stamens 2, with short connective appendage. Carpels on long stipe, usually many times longer than carpel .... 2. **Ruppia** L.

1. **Potamogeton** L.
Sp. pl. (1753) 126; Gen. pl. (1737) 33.

1. Leaves with well-developed sheath, invariably narrowly linear; plant totally submerged in water. Inflorescence elongated, invariably on slender peduncle and consisting of small, more or less interrupted verticels (subgenus *Coleogeton* Raunk.) .... 2.
+ Leaves variable, without or with short sheath; totally submerged in some species and partly so and partly floating in others. Inflorescence on more or less slender or often thick peduncle, closed, spicate, with more or less closely approximate flower verticels (subgenus *Potamogeton*) .... 7.
2. Lower part of leaf sheaths with connate edges and brown fringe; leaves and stem without strengthening tissue, hence quite soft. Verticels of inflorescence few (3–5), distant. Carpels with very short verrucose beak .... 5.
+ Leaf sheaths enlarged up to base, convoluted into tube, with white fringe; leaves and stem with fairly abundant strengthening tissue, more compact. Verticels of inflorescence generally 5–8 .... 3.
3. Carpels with distinct, more or less recurved beak (rostellum). Leaf tips more or less acuminate. Inflorescence generally with 5 unevenly interrupted verticels. Cauline leaves 5–15 cm long, 2.5 cm broad; stem strongly dichotomously branched, internodes 1.5–4 (10) cm long .... 3. **P. pectinatus** L.

+ Carpels with minute verrucose beak. Leaf tips rounded .... 4.
4. Leaf sheaths short, 3–4 (6) cm long, compact, highly developed, 0.5 cm broad, encompassing 3–4 (rarely 2) internodes, brownish-green; cauline leaves short, up to 10 cm long; basal leaves as long as cauline leaves; stem highly branched. Verticels of inflorescence numerous (generally 8), uniformly spaced .... 6. **P. vaginatus** Turcz.
+ Leaf sheaths narrow, long, up to 14 cm long, whitish-green lustrous; cauline leaves up to 12 cm long; basal leaves very long, up to 50 cm, with stiff edges; leaf tips recurved .... 4. **P. recurvatus** Hagstr.
5 (2). Plants large (up to 1 m tall); lower part of stem leafless, unbranched; leaves obtuse at tip, 12–20 cm long, 1–3 mm broad; sheath 2–3 cm long, 4–7 mm broad .... 2. **P. pamiricus** Baagöe.
+ Plants smaller (15–40 cm tall), branched from base; leaves cuspidate, acuminate or rarely, subobtuse .... 6.
6. Leaf tip with very short recurved cusp (resembling beak) ....
.... 5. **P. rostratus** Hagstr.
+ Leaf tip acuminate or subobtuse, without cusp ...... 1. **P. filiformis** Pers.
7 (1). All leaves submerged, slender and semi-transparent, usually similar in shape .... 8.
+ Leaves of 2 types: floating leaves petiolate, with rather thick, subcoriaceous, non-transparent broad blade, and submerged leaves linear, lanceolate or oblong .... 16.
8. Leaves lanceolate to suborbicular .... 9.
+ Leaves linear or linear-oblong .... 12.
9. Leaves more or less amplexicaul with rounded or cordate base, sessile .... 10.
+ Leaves not amplexicaul, petiolate or subsessile .... 11.
10. Leaves entire, with rounded or slightly amplexicaul base, elongated, ovate-lanceolate, up to 15 cm long, 1.5–2.5 cm broad, narrowing at tip into minute cap that soon ruptures and leaf becomes emarginate. Peduncles up to 20 cm long, many times longer than inflorescence .... 18. **P. praelongus** Wulf.
+ Leaves serrulate along margin, with cordate-amplexicaul base, relatively short and broad, oblong-ovate to suborbicular, 1.5–10 cm long, 1.5–3 cm broad; flat, obtuse or shortly acuminate at tip. Peduncles about 5 cm long, only 2–2.5 times longer than inflorescence ....
.... 17. **P. perfoliatus** L.
11. Leaves 5–8 cm long, 2 cm broad, on long petiole (up to 5 cm). Upper part of peduncles not enlarged .... 12. **P. malainus** Miq.
+ Leaves 8–20 cm long, 2–4.5 cm broad, sessile or narrowing into short and winged petiole. Upper part of peduncles enlarged, up to 7 cm long .... 11. **P. lucens** L.

12 (8). Leaves linear-oblong, with subparallel, dentate and highly undulate margins. Inflorescence with few and sparse flowers. Bases of carpels confluent, beak long (as long as carpel), usually falcate .... 7. **P. crispus** L.

+ Leaves linear, entire, margin not undulate. Inflorescence densely flowered. Carpels with short beak .... 13.

13. Leaves 2–8 cm long, 1–3 mm broad, with prominent veins at base, obtuse and with very short, barely visible cusp; stipules 1.5 cm long, broad, whitish, not connate .... 15. **P. obtusifolius** Mert. et Koch.

+ Leaves without prominent veins at base, acute, acuminulate or more rarely rounded, narrowing abruptly into short, distinct cusp .... 14.

14. Leaves 3.5–10 cm long, 1.3–2 mm broad, rigid, acuminate or acuminulate, with thick midrib and 12–16 fine lateral veins; stipules 2 cm long, pale, not connate .. 13. **P. manschuriensis** A. Benn.

+ Leaves slender, acute or rounded, narrowing into short cusp, with 3 veins .... 15.

15. Leaves apple-green, with midrib distinctly projecting downward, lateral veins weak and joining midrib at leaf tip at acute angle; stipules dense, light brown, connate for 2/3 length; stem branched in lower part, often simple towards top .... 16. **P. panormitanus** Biv.-Bern.

+ Leaves green with midrib not projecting downward but lateral veins joining midrib at tip almost at right angle; stipules transparent, not connate throughout length; stem simple or highly branched upward .... 19. **P. pusillus** L.

16 (7). Submerged leaves sessile, oblong or linear .... 17.

+ Submerged leaves long-petiolate, elliptical or ovate-lanceolate .... 18.

17. Plants small (up to 10 cm long). Submerged leaves narrowly linear, 2–3 cm long; floating leaves lanceolate or obovate, small (7–10 mm long, 2–3 mm broad) .... 10. **P. limosellifolius** Maxim.

+ Plants large (up to 100 cm tall). Submerged leaves linear-lanceolate, slender, translucent, with very fine teeth along margin, undulate, 4–8 cm long; floating leaves elliptical or oblong-elliptical, large (2–7 cm long, 1–3 cm broad) .... 9. **P. heterophyllus** Schreb.

18. Stem simple, 15–25 cm long; submerged leaves up to 10 cm long, with 11 veins; floating leaves up to 7 cm long, 3 cm broad, with 13–16 veins. Carpels convex ventrally, with central protuberance, semicircular on back with subacute middle and poorly visible lateral keels .... 8. **P. diginus** Wall.

+ Stem strongly branched; submerged leaves 10–30 cm long, with 7 veins; floating leaves 2–12 cm long, 0.5–3.5 cm broad, with 17–24 veins. Carpels with 3 keels on back, with considerably prominent midkeel .... 14. **P. nodosus** Poir.

Subgenus 1. C o l e o g e t o n Raunk.

1. **P. filiformis** Pers. Syn. pl. 1 (1805) 152; Baagöe in Vid. Medd. Dansk. Naturh. Foren (1903) 181; Aschers. and Graebn. in Engler, Pflanzenr. 31, IV, 11 (1907) 126; Hagstr. in Kungl. Sv. Vet. Ak. Handl. 55, 5 (1916) 14; id. in Hedin, S. Tibet, 6, 3 (1922) 96; Krylov, Fl. Zap. Sib. 1 (1927) 114; Juzepczuk in Fl. SSSR, 1 (1934) 236; Grubov, Konsp. fl. MNR (1955) 38; Fl. Kazakhst. 1 (1956) 90; Fl. Tadzh. 1 (1957) 92; Ikonnikov, Opred. rast. Pamira (1963) 40. —*P. marinus* auct. non L.: Ledeb. Fl. Ross. 4 (1853) 31. —**Ic.:** Pflanzenr. 31, IV, 11, fig. 28, C-E; Hagstr. l.c. (1916) fig. 3.

Described from Denmark. Type in London (Linn.) (?)

Lakes, brooks, stagnant water of ponds and tiny bogs, sedge meadows, sometimes in brackish water.

IA. Mongolia: *Cen. Khalkha* (saline puddle in Morin-Tologoi area, July 5, 1924—Pavl.; environs of Ikhe-Tukhum lake, source of Ata-Bulak—Mishikgun, June 1926—Zam.), *East. Mong.* (vicinity of Khailar town, in water, No. 754, June 19; vicinity of Sinbaerkhuyunchi town [Altyn-Emel'], No. 1031, June 29—1951, A.R. Lee (1959)), *Gobi-Alt.* (Bain-Tukhum area, depressions among solonchaks, in water, Aug. 4, 1931—Ik.-Gal.), *Ordos* (Bain-Nor, Sept. 2, 1884—Pot.).

IB. Kashgar: *Nor.* (Uchturfan, in garden, May 14, 1908—Divn.; Aksu, near Ai-Kul' village, Aug. 8, 1929—Pop.), *South.* (2 km east of Keriya town, in depressions in sand, No. 50, May 6, 1959—A.R. Lee (1959)), *East.* (Turfan, in water, 120 m, No. 6622, June 7; 3 km west of Turfan; same site, No. 6638, June 9; Toksun, in small ditch, 70 m, No. 7222, June 9; south of Pichan, in water, No. 6682, June 14; Yuili town [Chiglyk], near drying-up lake, No. 8602, Aug. 12—1958, Lee and Chou (A.R. Lee (1959)), *Nor.* (Maralbashi, in water, No. 7436, Sept. 6, 1958— Lee and Chou (A.R. Lee (1959)).

IIA. Junggar: *Tien Shan* (Kash river, 900 m, Sept. 6, 1878—A. Reg.; Sairam lake, July 23 [1878]—Fet.), *Jung. Gobi* (Savan district, Katszyvan, No. 1319, July 9, 1957—Kuan), *Balkh.- Alak.* (south of Dachen town [Chuguchak], Nos. 2866, 2876, Aug. 11, 1957—Kuan).

IIIB. Tibet: *Chang Tang* ("Northern Tibet, Upper Kum-Köl, freshwater lake, 3882 m, July 28, 1900"—Hagstr. in Hedin, l.c.).

IIIC. Pamir ["Eastern Pamir, Little Kara-kul, 3720 m, July 17, 1894; Tjakkeragil, freshwater lake, 3319 m, July 22, 1895; spring at Bulun-kul, 3405 m, July 23, 1895; lower Basik-kul, 3727 m, July 21 and 23, 1894"—Hagstr. in Hedin, l.c.).

General distribution: Aral-Casp., Jung.-Tarb., Tien Shan, East. Pam.; Arct. (Europ.), Europe (Nor.-west. and isolated in Central Volga region), Caucasus, Mid. Asia, West. Sib. (Altay), Far East (temperate), Nor. Mongolia (Fore Hubs., Hent.), China (Dunbei), Korean peninsula, Japan (nor.), North America.

Note. Species showing considerable variation but much less than *P. pectinatus* L. The variable habitats of plants depend mainly on variation in length of internodes and peduncles. Various investigators, especially Hagström (l.c.), have described a large number of varieties and races of this species. Hagström established 2 varieties for Cent. Asia: *P. filiformis* var. *tibetanus* Hagström and var. *linipes* Hagström (Bot. notiser, 1905, 142). Var. *tibetanus* ("ofre Kum-köl, N. Tibet") differs from type specimen in much less height, strong branching and very short internodes (0.5–5 cm); in var. *linipes* Hagstr. (Tibet) uppermost internode 10–15 cm long and peduncles 25–30 cm long.

2. **P. pamiricus** Baagöe in Vid. Medd. Dansk. Naturh. Foren (1903) 182; Aschers and Graebn. in Engler, Pflanzenr. 31, IV, 11 (1907) 127; in Hagstr. in Kungl. Sv. Vet. Ak. Handl. 55, 5 (1916) 25; O. and B. Fedtsch. in Tr. Glavn. bot. sada, 38, 1 (1924) 53; Juzepczuk in Fl. SSSR, 1 (1934) 237; Fl. Kirgiz. 1 (1952) 86; Fl. Kazakhst. 1 (1956) 91; Fl. Tadzh. 1 (1957) 92; Ikonnikov, Opred. rast. Pamira (1963) 38.

Described from East. Pamir (Kara-Kul' lake). Type in Copenhagen.

Lakes and small water reservoirs, bog pools, hot springs, upper belt of mountains.

IIIB. Tibet: *Chang Tang* ("Tibet occ., 3600–4500 m, Thomson"—Hagstr. l.c.).
IIIC. Pamir (5–6 km south of Upal oasis, along road to Tashkurgan, discharge of springs, in water, June 11, 1959—Yun.).
General distribution: Cent. Tien Shan, East. Pam.; Mid. Asia (Pamiro-Alay), Himalayas (Kashmir).

3. **P. pectinatus** L. Sp. pl. (1753) 127; Ledeb. Fl. Ross. 4 (1853) 30; Hook. f. Fl. Brit. India, 6 (1893) 567; Hemsley, Fl. Tibet (1902) 200; Forbes and Hemsley, Index Fl. Sin. 3 (1903) 195; Baagöe in Vid. Medd. Dansk. Naturh. Foren (1903) 181; Aschers and Graebn. in Engler, Pflanzenr. 31, IV, 11 (1907) 121; Danguy in Bull. Mus. nat. hist. natur. 20 (1914) 143; Hagstr. in Kungl. Sv. Vet. Ak. Handl. 55, 5 (1916) 39; id. in Hedin, S. Tibet, 6, 3 (1922) 96; Krylov, Fl. Zap. Sib. 1 (1927) 113; Pampanini, Fl. Carac. (1930) 69; Juzepczuk in Fl. SSSR, 1 (1934) 236; Kitag. Lin. Fl. Mansh. (1939) 53; Persson in Bot. notiser (1938) 273; Hao in Engler's Bot. Jahrb. 68 (1938) 578; Norlindh, Fl. mong. Steppe, 1 (1949) 43; Fl. Kirgiz. 1 (1952) 86; Grubov, Konsp. fl. MNR (1955) 59; Fl. Kazakhst. 1 (1956) 91; Fl. Tadzh. 1 (1957) 93; Ikonnikov, Opred. rast. Pamira (1963) 40; Hanelt et Davažamc in Feddes Repert. 70, 1–3 (1965) 12. —**Ic.:** Pflanzenr. 31, IV, 11, fig. 28, A–B; Hagstr. l.c. (1916), fig. 18.

Described from Europe. Type in London (Linn.).

Lakes and rivers at shallow depths, meanders and streams, sometimes in reservoirs with brackish water.

IA. Mongolia: *Mong. Alt.* (Dain-Gol lake, discharge of Kutan, July 6, 1906—Sap.; nor.-east. end of Tonkhil'-Nur lake, in water on bank of freshwater pond, Sept. 7, 1948—Grub.), *Cis-Hing.* (near Yaksha railway station, lake, Aug. 19, 1962—Litw.; Arshan, in flowing water of hot springs, No. 294, June 15, 1950—Chang), *Cent. Khalkha* (midcourse of Kerulen, in flowing rivers along valley, near Batur-Chzhonon-Tszasaka camp; Khoshun-Batun, Dzhong-Dzolon, 1899—Pal.; Kerulen river valley, puddle, Aug. 10, 1924—Lis.; Olon-Nor lake, silty soil, Sept. 17, 1925—Gus.), *East. Mong.* (Bain-Nor lake [southern end of Buir-Nur lake], in water along bank, June 19, 1899—Pot.; "Doyen in laculo, fruct., Aug. 17, 1934, Eriksson, No. 805"—Norlindh, l.c.), *Val. Lakes* ("Bajanchongor: im SO-Teil des Buncagan-nur, Brackwasser, ca. 2 m Wassertiefe [2191]"—Hanelt et Davažamc, l.c.), *East. Gobi* ("Khujirtu-gol, Camp VIII, in rivulo, flor. June 28, 1927, No. 1176 et Khohnin-chaghan-chölogoe, Camp XI, ster. Aug. 1, 1927, 1323, Hummel"—Norlindh, l.c.), *Alash. Gobi* (Churgi-Bulyk, Aug. 10, 1873—Przew.;

Khara-Sukhai, Edzin river, July 20, 1886—Pot.; Bayan-Khoto, Tengeri sand, Bagedabusu, July 12, 1958—Petr.; "Sogho-nor, in aqua, ster., Sept. 8, 1927, No. 1759, Hummel; ibid, Camp VII, flor., July 13, 1928, Nos. 7052, 7053; Bayan-Bogdo, Camp Wen-tsun-hai-tze, flor., June 22, 1929, No. 7620—Söderbom; prope flumen Edsengol, in laculo, ca. 1 km ad occid. versus a Camp LVIII, flor., May 2.7 1930, No. 1859, Bohlin"—Norlindh, l.c.), *Ordos* (Huang He river valley, Aug. 4, 1871—Przew.; 75 km south of Dzhasak town, Tautykhaitsy lake, in water, Aug. 17, 1957—Petr.), *Khesi* marshy bank of Edzin-Gol river near Gaotai, June 6 and 20—1886, Pot.).

IB. Kashgar: *Nor.* ("Lower Tarim, ca. 830, 1900, Hedin"—Hagstr. in Hedin, l.c.), *Lob-Nor* ("Mapick-köl, a part of Kara-koschum, 816 m, June 23, 1900, Hedin"—Hagstr. in Hedin, l.c.).

IIA. Junggar: *Cis-Alt.* ("Altai entre la vallee de la riviere Koun et l'Irtich, alt. 1470 m, Aug. 21, 1896, Chaffanjon"—Danguy, l.c.), *Tien Shan* ("Köl, in a small mountain valley with a swamp ca. 2700 m, Aug. 7, 1932"—Persson, l.c.), *Jung. Gobi* (lower Manas river, running backwater overgrown with reeds, inundated site, June 20, 1957—Yun.).

IIIA. Qinghai: *Nanshan* (Tetung river, near Khara river estuary, marshy banks, 3450 m, June 1886—Pot.; "Kokonor: im salzigen Wasser des Sees Da-lian-nor, 3500 m"—Hao, l.c.).

IIIB. Tibet: *Weitzan* (Alak-Nor-gol river, in springs, common, Aug. 11, 1884—Przew.).

General distribution: Aral-Casp., Fore Balkh., Jung.-Tarb., Tien Shan, East. Pamir; cosmopolitan.

Note. *P. pectinatus* represents one of the more widely distributed and extremely polymorphous species of the genus. Undoubtedly, this is a composite species calling for thorough investigation. Various investigators have described for this species numerous varieties and races with different taxonomic relations.

Var. *Coronatus* Hagstr. l.c. (Muntjokk-ott, June 23, Mapick-köl) reported from Central Asia differs in numerous (5–8) verticels of inflorescence, small (3 × 2 mm) carpels and narrow (0.5–1 mm) leaves.

Bennett [J. Bot. (London) 1894, 203–204] described var. *mongolicus* A. Benn. characterised by trichoid stems, bright green setaceous 1-veined leaves, very long (up to 25 cm) peduncle, short inflorescence with 3 verticels and small carpels. This variety is represented in the Herbarium of the Komarov Botanical Institute by N.M. Przewalsky's 1871 collection from Huang He river valley; a duplicate of this specimen served as type for establishing this variety.

4. P. recurvatus Hagstr. in Kungl. Sv. Vet. Ak. Handl. 55, 5 (1916) 37; id. in Hedin, S. Tibet, 6, 3 (1922) 96. —Ic.: Hagstr. l.c. (1916) fig. 14.

In clean water ponds.

Described from Pamir. Type in London.

IIIB. Tibet: *Chang Tang* ("S.E. Tibet, small somewhat brackish lake near Camp XIV, 4968 m, Aug. 28, 1896 [ster.]; Tibet, Camp LXVI, in a lake, Aug. 26, 1901; Camp LXXVIII and Camp LXXIX, Naktsongtso, a little freshwater lake, 4636 m, Sept. 11–22, 1901; lake, 4674 m, Sept. 12, 1901; S.W. Tibet, on way between Camp CCIII [Dara-sumkor], 4931 m and Camp CCIV [Bak-gyäorap], 4870 m, northern foot of Himalayas, July 16, 1907 [ster.]"—Hagstr. l.c.).

IIIC. Pamir ("a specimen probably from Kuen-Luen, but labelled by J. Baägoe as collected by Ove Paulsen in Pamir"—Hagstr. l.c. [1916]—typus!).

General distribution: endemic.

5. P. rostratus Hagstr. in Kungl. Sv. Vet. Ak. Handl. 55, 5 (1916) 27; id. in Hedin, S. Tibet, 6, 3 (1922) 96; Grubov, Konsp. fl. MNR (1955) 59.

Described from East. Tien Shan. Type in Stockholm (S).
Lakes, ponds, meanders, marshes.

IA. Mongolia: *East. Mong.* (in Naryn-gol brook [Tsagan-balgasu], June 6, 1831—Ladyzh.), *Bas. Lakes* (near Khirgis-Nur lake, Aug. 5, 1879—Pot.), *Val. Lakes* (Orok Nor, in shallow lagoons at 1110 m, No. 297, 1925—Chaney), *Gobi-Alt.* (Bain-Tukhum area, depressions among meadow solonchaks, in water, Aug. 29, 1931—Ik.-Gal.; Tsagan-Gol river, in running water, July 27, 1948—Grub.), *Alash.-Gobi* (Dyn'yuanin oasis [Bayan-Khoto], in springs, on muddy silt, March 20, 1908—Czet.).

IB. Kashgar: *Nor.* (in Bai town region, Kee village, in water, 1580 m, No. 8153, Sept. 1; 3 km north of Aksu, pond in solonchak wasteland, 1100 m, No. 8971, Sept. 30—1958, Lee and Chu), *West.* (west of Upal village, on water, No. 276, June 11, 1959—A.R. Lee (1959)), *South.* (Kuenluen, prov. Kotan, Lake Kirik-Kiöl, July 13–14 and Oitash, down to foot of Buschia glacier [northern side of Kuenluen] Aug. 27—1856, Schlagintweit), *East.* (Khami town, No. 449, May 21, 1957—Kuan; nor.-east of Toksun, on water, 300 m, No. 7337, June 19, 1958—Lee and Chu).

IIA. Junggar: *Jung. Alt.* (Toli town region, on water, No. 2448, Aug. 4, 1957—Kuan), *Tien Shan* (B. Yuldus, on marsh near water, 2460 m, No. 6483, Aug. 10, 1958—Lee and Chu (A.R. Lee (1959)); intermontane basin of B. Yuldus, 30–35 km south-west of Bain-Bulak settlement, nor. margin of marshy floor of basin, on water, Aug. 10, 1958—Yun.), *Jung. Gobi* ("desertum a Thianschan boream versus, 1877, No. 6039—Pot.", typus!—Hagstr. l.c. [1916]; Ubchugiin-Gol sources, in water of spring, at discharge, Sept. 9, 1948—Grub.; between Tien Shan Laoba in Myaoergou, No. 2399, Aug. 3, 1957—Kuan).

IIIA. Qinghai: *Nan Shan* (Kuku-Nor lake, south. bank, 1901—Lad.).

IIIB. Tibet: *Chang Tang* ("northern Tibet, Temirlik, Camp VII, 2961 m, July 10, 1900, Hedin"—Hagstr.), *Weitzan* (Dzhagyn-Gol river, 4050 m, in grasslands and small ponds on marshy banks of river, July 3, 1900—Lad.).

General distribution: Nor. Mong. (Hang.), China North-west).

6. **P. vaginatus** Turcz. in Bull. Soc. Natur. Moscou, 27 (1854) 66; ej. Fl. baic.-dah. 2 (1856) 162; Hagstr. in Kungl. Sv. Vet. Ak. Handl. 55, 5 (1916) 32; Juzepczuk in Fl. SSSR, 1 (1934) 238; Grubov, Konsp. fl. MNR (1955) 59; Fl. Kazakhst. 1 (1956) 91. —*P. pectinatus* var. *vaginatus* (Turcz.) Aschers. et Graebn. Synops. 1 (1897) 351; id. in Engler. Pflanzenr. 31, IV, 11 (1907) 124.

Described from East. Siberia (Transbaikal, Selengin lake). Type in Leningrad. Plate V, fig. 2.

Lakes, ponds, solonchak marshes.

IA. Mongolia: *Mong. Alt.* (nor.-east. end of Tonkhil'-Nur lake, on water near bank of freshwater pond, Sept. 7, 1948—Grub.), *Cent. Khalkha* (Ugei-Nor, vicinity of bank, June 17, 1893—Klem.; ponds around Tszagaste river, Sept. 23, 1927—Gus.), *Val. Lakes* (Kolobolchi Nor, in lake at 1230 m, sandy bottom, in water 0.6–1.2 m deep, No. 289, 1925—Chaney).

IB. Kashgar: *East.* (near Bagrashkul' lake, 1000 m, No. 6202, July 26, 1958—Lee and Chu (A.R. Lee (1959)).

IIA. Junggar: *Jung. Gobi* (along Karamai-Urkho highway, 20 km south-south-east of bridge on Darbuty river, Manas river current, backwaters overgrown with reeds, in clean water, June 20, 1957—Yun.).

General distribution: Aral-Casp., Fore Balkh.; Arct. (Yenisei lowland), Europe (stray sites on bank of Gulf of Bothnia and in Karelian ASSR). East. Sib. (Gis Angara area and Transbaikal), Nor. Mong. (Hang., Mong.-Daur.), Nor. America (north).

Subgenus 2. P o t a m o g e t o n

7. **P. crispus** L. Sp. pl. (1753) 126; Hook. f. Fl. Brit. India, 6 (1893) 566; Forbes and Hemsley, Index Fl. Sin. 3 (1903) 193; Aschers. and Graebn. in Engler, Pflanzenr. 31, IV, 11 (1907) 97; Hagstr. in Kungl. Sv. Vet. Ak. Handl. 55, 5 (1916) 58; Krylov, Fl. Zap. Sib. 1 (1927) 110; Pampanini, Fl. Carac. (1930) 69; Juzepczuk in Fl. SSSR, 1 (1934) 240; Kitag. Lin. Fl. Mansh. (1939) 52; Norlindh, Fl. mong. steppe, 1 (1949) 44; Fl. Kirgiz. 1 (1952) 86; Fl. Kazakhst. 1 (1956) 92; Fl. Tadzh. 1 (1957) 94; Ikonnikov, Opred. rast. Pamira (1963) 41. —*P. serrulatus* Schrad. et Opiz in Flora, 5 (1822) 267. —**Ic.:** Pflanzenr. 31, IV, 11, fig. 23, A–C; Hagstr. l.c. figs., 21, 22; Fl. SSSR, 1 Plate XII, fig. 6a–c.

Described from Europe. Type in London (Linn.).

Stagnant or slow-flowing water of ponds, backwaters, lakes, ditches, occasionally in rivers.

IA. Mongolia: *Cent. Khalkha* (Kerulen valley, very small lake, Aug. 10, 1924—Lis.), *Alash. Gobi* (Dyn'yuan'in oasis [Boyan-Khoto], June 27, 1908—Czet.; "Bayan-Bogdo Camp, Wentsun-hai-tze, ster., June 22, 1929—Söderbom, No. 7619"—Norlindh, l.c.).

IB. Kashgar: *East.* (Turfan, Nov. 11, 1879—A. Reg.).

General distribution: Aral-Casp., Fore Balkh., Cent. Tien Shan, East. Pam.; Europe, Mediterranean, Balk.-Asia Minor, Fore Asia, Caucasus, Mid. Asia, West. Sib. (south. Altay), East. Sib. (Transbaikal), Far East (south), Nor. Mongolia (?), China (Dunbei, North, North-west, East, South-west), Korean peninsula, Japan, India, North America, Africa, Australia.

8. **P. diginus** Wall. Cat. (1828) 5177; Juzepczuk in Fl. SSSR, 1 (1934) 253. —*P. oblongus* Viv. Ann. bot. 1, 2 (1802) 102; Hook. f. Fl. Brit. India, 6 (1893) 566. —*P. polygonifolius* auct. non Pourr.: Forbes and Hemsley, Index Fl. Sin. 3 (1903) 196; Aschers and Graebn. in Engler, Pflanzenr. 31, IV, 11 (1907) 65, p.p.; Hagstr. in Kungl. Sv. Vet. Ak. Handl. 55, 5 (1916) 175, p.p.

Described from Himalayas (Nepal). Type in London (K).

Running water.

IA. Mongolia: *Ordos* (Dzhasygen, Aug. 30, 1884—Pot.), *Khesi* (between Fuiitin and Gaotai towns, June 6 and vicinity of Gaotai town, June 25—1886, Pot.).

General distribution: Far East (south), China (Dunbei, Nor. Cent., East, South-west, Taiwan), Korean peninsula, Japan, Indo-Mal.

9. **P. heterophllus** Schreb. Spicil. Fl. Lips. (1771) 21; Juzepczuk in Fl. SSSR, 1 (1934) 256; Kitag. Lin. Fl. Mansh. (1939) 52; Grubov, Konsp. fl. MNR (1955) 59; Fl. Kazakhst. 1 (1956) 95; Fl. Tadzh. 1 (1957) 99; Ikonnikov, Opred. rast. Pamira (1963) 41. —*P. gramineus* L. Sp. pl. (1753) 127, p.p.; Aschers. and Graebn. in Engler, Pflanzenr. 31, IV, 11 (1907) 84; Hagstr. in Kungl. Sv. Vet. Ak. Handl. 55, 5 (1916) 204; Krylov, Fl. Zap. Sib. 1 (1927) 109; Pampanini, Fl. Carac. (1930) 69. —**Ic.:** Pflanzenr. 31, IV, 11, fig. 20, A–C.

Described from Europe. Type in Leningrad (?).

Lakes and streams with poorly running water.

IA. Mongolia: *Cent. Khalkha* (Kerulen valley, very small lake, Aug. 10, 1924—Lis.; Ubur-Dzhargalante river source, very small lake in valley, Sept, 13, 1925—Krasch.), *East. Mong.* (near Chingiskhan railway station, meadow lake, July 23, 1903—Litw.; in Sinbaerkhuyunchi district, on water, No. 1071, June 30, 1951—A.R. Lee et al.), *Bas. Lakes* (near Khirgis-Nur lake on Tatkhen-Teli river, Dzabkhyna tributary, Aug. 11, 1879—Pot.). *Ordos* (Taitukhai, Aug. 29; Dzhasygen-Qaidam area, Aug. 30—1884, Pot.; 75 km south of Dzhasak, on water, Aug. 17, 1957—Petr.).

IIIB. Tibet: *Chang Tang* ("Lago Pangong: Sciusciul, Scharma-la, 4400 m"—Pampanini, l.c.).

General distribution: Fore Balkh., Jung.-Tarb., East. Pam.; Arct., Europe, Caucasus, West. and East. Siberia, Far East, Nor. Mongolia (Hent., Hang., Mong.-Daur.), China (Dunbei), Korean peninsula, Japan, North America.

Note. Polymorphous species with several races and varieties. Many investigators have reported a large number of hybrids in Europe.

10. **P. limesellifolius** Maxim. ex Korsh. in Acta Horti Petrop. 12 (1892) 393; Forbes and Hemsley, Index Fl. Sin. 3 (1903) 195; Aschers. and Graebn. in Engler, Pflanzenr. 31, IV, 11 (1907) 50; Hagstr. in Kungl. Sv. Vet. Ak. Handl. 55, 5 (1916) 107; Juzepczuk in Fl. SSSR, 1 (1934) 246. —**Ic.:** Hagstr. l.c. fig. 43; Fl. SSSR, 1, Plate XII, fig. 13a–c.

Described from Far East (Zeisk-Bureinsk interfluve). Type in Leningrad.

In meanders.

IA. Mongolia: *East. Mong.* (around Khailar town, in flowing water, No. 2153, Aug. 29, 1951—A.R. Lee (1959)).

General distribution: Far East, China (Dunbei), Korean peninsula, Japan.

Note. While describing *P. limosellifolius*, S.I. Korshinskii erred by including in the specimens studied one collected by R. Maack "... Ussuri medium, prope Buldshi ..." which is authentic for *P. cristatus* Regel et Maack. Although at first sight these species are externally similar, they differ sharply in carpels. Carpel of *P. cristatus* has ventrally pectinate, long-toothed keel and 2 long corniculate protuberances at base; carpel of *P. limosellifolius* has no such appendage.

11. **P. lucens** L. Sp. pl. (1753) 126; Hook. f. Fl. Brit. India, 6 (1893) 567; Forbes and Hemsley, Index Fl. Sin. 3 (1903) 195; Aschers. and Graebn. in Engler, Pflanzenr. 31, IV, 11 (1907) 76; Hagstr. in Kungl. Sv. Vet. Ak. Handl. 55, 5 (1916) 233; Krylov, Fl. Zap. Sib. 1 (1927) 108; Juzepczuk in Fl. SSSR, 1 (1934) 257; Kitag. Lin. Fl. Mansh. (1939) 52; Fl. Kirgiz. 1 (1952) 89; Grubov, Konsp. fl. MNR (1955) 59; Fl. Kazakhst. 1 (1956) 96; Fl. Tadzh. 1 (1957) 99; Ikonnikov, Opred. rast. Pamira (1963) 40. —**Ic.:** Pflanzenr. 31, IV, 11, fig. 18, A–D; Flora SSSR, 1, Plate XII, fig. 23a–c.

Described from Europe. Type in London (Linn.).

Rivers, streams, lakes, predominantly in stagnant water.

IA. Mongolia: *Bas. Lakes* (Khirgis-Nur lake, Chon-Kharikha river and Khara-Nur lake, Aug. 12, 1879—Pot.).

IB. Kashgar: *Nor.* (south-east of Yuili [Chiglyk], on water, No. 8538, Aug. 5, 1958—Lee and Chu).

IIA. Junggar: *Jung. Gobi* (Savan district, Poatai [Shamyn'tsza], No. 1595, June 29, 1957—Kuan), *Balkh.-Alak.* (south of Dachen [Chuguchak] town, No. 2862, Aug. 11, 1957—Kuan).

General distribution: Aral-Casp., Fore Balkh., Jung.-Tarb., Cent. Tien Shan, East. Pamir; Europe, Mediterranean; Balk.-Asia Minor, Fore Asia, Caucasus, Mid. Asia, West. Sib., East. Sib. (south-west.), China (Dunbei, nor.), Himalayas (Kashmir), Japan, North America.

Note. Differs from closely related *P. malainus* Miq. in sessile leaves (in *P. malainus*, leaves on long slender petiole) as well as in enlarged upper part of peduncle.

12. **P. malainus** Miq. Ill. Fl. Arch. Indien (1871) 46; Aschers. and Graebn. in Engler, Pflanzenr. 31, IV, 11 (1907) 83; Hagstr. in Kungl. Sv. Vet. Ak. Handl. 55, 5 (1916) 248; Juzepczuk in Fl. SSSR, 1 (1934) 258; Kitag. Lin. Fl. Mansh. (1939) 52; Fl. Kazakhst. 1 (1956) 96. —*P. mucronatus* auct. non Schrad.: Hook. f. Fl. Brit. India, 6 (1893) 567. —**Ic.:** Pflanzenr. 31, IV, 11, fig. 18, E, F; Hagstr. l.c. fig. 115, A–E; Fl. SSSR, 1, Plate XII, fig. 24a–c.

Described from Malay archipelago. Type in Leyden.

In rivers.

IA. Mongolia: *Ordos* (in Huang He river valley, on water, Aug. 4, 1871—Przew.).
IB. Kashgar: *Nor.* (near Karashar town, in brook, Aug, 25, 1929—Pop.).
General distribution: Far East (south.), China (Dunbei, South-west.), Korean peninsula, Japan, Indo-Mal.

13. **P. manschuriensis** A. Benn. in J. Bot. (London) 42 (1904) 76; Juzepczuk in Fl. SSSR, 1 (1934) 243; Kitag. Lin. Fl. Mansh. (1939) 53.

Described from China (Sungari river). Type in Leningrad.

Ponds, meanders and along banks of brooks.

IA. Mongolia: *East. Mong.* (near Chingiskhan railway station, lake, July 23, 1903—Litw.).
General distribution: Far East (south.), China (Dunbei).
Note. Only the single incomplete specimen mentioned above is known for Cent. Asia and its placement under *P. manschuriensis* raises some doubt. While describing *P. manschuriensis*, Bennett examined the specimen but did not identify it as *P. manschuriensis*, noting rather its defective condition. Its label is available in the herbarium sheet. Nevertheless, the affinity of this specimen to group Oxyphylli Hagstr. is beyond doubt and we have as yet little justification to identify it with any other species of this group. Judging from the distribution range of *P. manschuriensis*, its occurrence in East. Mongolia is entirely probable, which cannot be said for other species of this group.

14. **P. nodosus** Poir. apud Lam. Encycl. mét. bot. suppl. 4 (1816) 535; Hagstr. in Kungl. Sv. Vet. Ak. Handl. 55, 5 (1916) 183; Juzepczuk in Fl. SSSR, 1 (1934) 254; Fl. Kazakhst. 1 (1956) 95; Fl. Tadzh. 1 (1957) 98. —*P. americanus* Cham. et Schlecht. in Linnaea, 2 (1827) 226. —*P. fluitans* subsp. *americanus* Aschers. et Graebn. in Engler, Pflanzenr. 31, IV, 11 (1907) 60. —**Ic.:** Hagstr. l.c. fig. 95; Fl. SSSR, 1, Plate XII, fig. 20.

Described from Canary islands. Type in Monpel.

In rivers, influent lakes.

IA. Mongolia: *East. Mong.* (? near Chingiskhan railway station, influent lake, July 23, 1903—Litw.).

IB. Kashgar: *Nor.* (near Aksu town, around Aikul' village, Aug. 8, 1929—Pop.), *East.* (Kuruktag mountain range, Kurla river gorge, Bash-akin picket, rapids, Aug. 24, 1929—Pop.; Shizinko in Turfan, 220 m, No. 6623, June 8, 1958—Lee and Chu (A.R. Lee (1959)).

IIA. Junggar: *Jung. Gobi* (Usu-San'tszyao-Chzhuan, No. 1054, June 25, 1957—Kuan), *Balkh.-Alak.* (south of Dachen town [Chuguchak], No. 2868, Aug. 11, 1957—Kuan).

General distribution: Aral-Casp., Fore Balkh.; Europe, Fore Asia, Caucasus, Mid. Asia, West. Sib. (south-west.), India, North and South America, Africa.

15. **P. obtusifolius** Mert. et Koch, Deutschl. Fl. ed. 3, 1 (1823) 855; Aschers. and Graebn. in Engler, Pflanzenr. 31, IV, 11 (1907) 108; Hagstr. in Kungl. Sv. Vet. Ak. Handl. 55, 5 (1916) 115; Krylov, Fl. Zap. Sib. 1 (1927) 111; Juzepczuk in Fl. SSSR, 1 (1934) 246; Kitag. Lin. Fl. Mansh. (1939) 53; Fl. Kirgiz. 1 (1952) 89; Fl. Kazakhst. 1 (1956) 94. —Ic.: Pflanzenr. 31, IV, 11, fig. 27, A–D; Hagstr. l.c. fig. 51; Fl. SSSR, 1, Plate XII, fig. 14.

Described from Europe. Type in Berlin.

In standing water in ditches, ponds, small lakes, river backwaters.

IA. Mongolia: *Ordos* (Taitukhai, Aug. 29, 1884—Pot.).

IIA. Junggar: *Tien Shan* (Bogdo mountain, 1650 m, Aug. 15, 1878—A. Reg.).

General distribution: Aral-Casp., Fore Balkh.; Europe, Fore Asia (west. Iran), Mid. Asia (West. Tien Shan), West. Sib. (south.), East. Sib. (south-west.), China (Dunbei), North America.

16. **P. panormitanus** Biv.-Bern. Nuove piante ined. Palermo (1838) 6; Hagstr. in Kungl. Sv. Vet. Ak. Handl. 55, 5 (1916) 98; Juzepczuk in Fl. SSSR, 1 (1934) 246; Norlindh, Fl. mong. steppe, 1 (1949) 45. —*P. pusillus* proles *panormitanus* (Biv.-Bern.) A. Benn. in J. Bot. (London) 19 (1881) 67; Aschers and Graebn. in Engler, Pflanzenr. 31, IV, 11 (1907) 116. —*P. noltei* A. Benn. in J. Bot. (London) 28 (1890) 30. —Ic.: Hagstr. l.c. figs. 38, 39; Fl. SSSR, 1, Plate XII, fig. 12.

Described from Europe (Sicily). Type in Palermo.

Lakes and rivers.

IA. Mongolia: *East. Mong.* ("20 km ad orient. vers. a Doyen, in laculo silvestri flor. et fruct., July 31, 1924—Eriksson, No. 58"—Norlindh, l.c.), *East. Gobi* ("Hailutain-gol, Camp Norinii XXII, distr. Dunda-gung, fruct. Aug. 8, 1927, No. 682—Söderbom"—Norlindh, l.c.).

General distribution: Europe, Fore Asia (Afghanistan), East. Sib. (midcourse of Yenisei), Far East (Kamchatka), China (Nor. North-west.), North America, Africa.

17. **P. perfoliatus** L. Sp. pl. (1753) 126; Henders. and Hume, Lahore to Jarkand (1873) 337; Hook. f. Fl. Brit. India, 6 (1893) 566; Forbes and Hemsley, Index Fl. Sin. 3 (1903) 196; Aschers. and Graebn. in Engler, Pflanzenr. 31, IV, 11 (1907) 92; Danguy in Bull. Mus. nat. hist. natur. 17 (1911) 451; Hagstr. in Kungl. Sv. Vet. Ak. Handl. 55, 5 (1916) 254; id. in Hedin, S. Tibet, 6, 3 (1922) 95; Krylov, Fl. Zap. Sib. 1 (1927) 106; Pampanini, Fl. Carac. (1930) 69; Juzepczuk in Fl. SSSR, 1 (1934) 260; Persson in Bot. notiser (1938) 273; Kitag. Lin. Fl. Mansh. (1939) 53; Norlindh, Fl. mong. steppe, 1 (1949) 46; Fl. Kirgiz. 1 (1952) 89; Grubov, Konsp. fl. MNR (1955) 59; Fl. Kazakhst. 1 (1956) 97; Chen and Chou, Rast. pokrov r. Sulekhe (1957) 91; Fl. Tadzh. 1 (1957) 100;

Ikonnikov, Opred. rast. Pamira (1963) 40. —Ic.: Pflanzenr. 31, IV, 11, fig. 21; Hagstr. l.c. (1916) figs. 117, 118A; Fl. SSSR, 1, Plate XII, fig. 26a–c.

Described from Europe. Type in London (Linn.).

In stagnant waters of lakes, ponds, ditches and slow-flowing rivers, sometimes at fairly deep level. Frequently forms considerable patches. One of the most widely distributed pondweeds.

IA. Mongolia: *Mong. Alt.* (Dain-Gol lake, at bottom in backwaters, July 28, 1908—Sap.), *Cis-Hing.* (near Yaksha railway station, meadow lake, Aug. 19, 1902—Litw.), *Cent. Khalkha* (in Ugei-Nur lake, near bank, June 17, 1893—Klem.; midcourse of Kerulen, in river currents, near Batur-Chzhonon-Tszasaka station, 1889; Khoshun-Batur-Chzhon-Dzolon and Ulugui steppe river, 1899—Pal.), *East. Mong.* (near Chingiskhan railway station, lake, July 23, 1903—Litw.; Buir-Nor lake, south-west. bank, Aug. 29; Khalkhin-Gol river, Symbur area, Sept. 1—1928, Tug.; around Khailar town, on water, No. 752, June 19; same site, No. 2152, Aug. 29—1951, A.R. Lee (1959); "Doyen, in aqua stagnante, flor. et fruct., Aug. 5, 1926, Eriksson, No. 246"—Norlindh, l.c.), *Bas. Lakes* (around Khirgis-Nur lake, Aug. 2; same site, Dzabkhyn river, Aug. 2; Chon-Kharikha river, Aug. 12—1879, Pot.), *Val. Lakes* (Orok-Nor, lagoons at 1100 m, No. 302, 1925—Chaney; Orok-Nur lake, wet lacustrine lagoons and solonetzes, Aug. 4, 1926—Tug.), *Alash. Gobi* (Edzin-Gol river, June 25, 1886—Pot.; Edzin-Gol river valley, upper Ontsin-Gol branch, Bukhan-khub area, July 13, 1926—Glag.; "prope fl. Edsen-gol, in laculo ca. 1 km occid. versus a Camp 58, flor., 27 Maj 1930, No. 1858, Bohlin"—Norlindh, l.c.), *Khesi* (between Fuiitin and Gaotai towns, June 6, 1886—Pot.; "Lac de Je-Jue-Tsiuan, alt. 1000 m, environs de Cha-tcheou, June 4, 1908, Vaillant"—Danguy, l.c.; Sulekhe river—Chen and Chou, l.c.).

IB. Kashgar: *Nor., East.* (near Bagrashkul' lake, on water, 1100 m, No. 6203, June 26; east of Yuili [Chiglyk], on water, No. 8539, Aug. 5—1958, Lee and Chu (A.R. Lee (1959)).

IC. Qaidam: *Montane* (Kurlyk-Nor lake, south., freshwater, silty floor, 2580 m, June 29; Khara-Usu river, 2580 m, Aug. 2, 1901—Lad.).

IIA. Junggar: *Balkh.-Alak.* (south of Dachen [Chuguchak], No. 2869, Aug. 11, 1957—Kuan).

IIIA. Qinghai: *Nanshan* (Kuku-Nor lake, south. bank, 1901—Lad.).

General distribution: Aral-Casp., Fore Balkh., Jung,-Tarb., Cent. Tien Shan, East. Pamir; Arct., Europe, Caucasus, Mid. Asia, West. and East. Sib., Far East, Nor. Mong. (Fore Hubs., Hent., Hang.), China (Dunbei, nor.), Himalayas (Kashmir), Korean peninsula, Japan, North America, Africa (nor.), Australia.

18. **P. praelongus** Wulf. in Roem. Arch. Bot. 3, 3 (1805) 331; Aschers. and Graebn. in Engler, Pflanzenr. 31, IV, 11 (1907) 96; Hagstr. in Kungl. Sv. Vet. Ak. Handl. 55, 5 (1916) 250; Krylov, Fl. Zap. Sib. 1 (1927) 107; Juzepczuk in Fl. SSSR, 1 (1934) 259; Grubov, Konsp. fl. MNR (1955) 59; Fl. Kazakhst. 1 (1956) 97. —Ic.: Pflanzenr. 31, IV, 11, fig. 22; Hagstr. l.c. figs. 116A, 117P, 118C; Fl. SSSR, 1, Plate XII, fig. 25a, b.

Described from Europe (Poland). Type not known.

In lakes and rivers (common in fairly deep sites).

IA. Mongolia: *Cent. Khalkha* (Ugei-Nor, near bank, June 17, 1893—Klem).

IIA. Junggar: *Jung. Gobi* (in Barbagai region, around solonchak, 500 m, No. 3238, Sept. 20, 1956—Ching).

General distribution: Arct., Europe, Caucasus, West. and East. Sib., Far East, China (Dunbei), Japan, North America.

19. **P. pusillus** L. Sp. pl. (1753) 127; Franch. Pl. David. 1 (1884) 315; Hook. f. Fl. Brit. India, 6 (1893) 567; Forbes and Hemsley, Index Fl. Sin. 3 (1903) 196; Aschers. and Graebn. in Engler, Pflanzenr. 31, IV, 11 (1907) 113; Danguy in Bull. Mus. nat. hist. natur. 20 (1914) 143; Hagstr. in Kungl. Sv. Vet. Ak. Handl. 55, 5 (1916) 121; Krylov, Fl. Zap. Sib. 1 (1927) 112; Juzepczuk in Fl. SSSR, 1 (1934) 247; Kitag. Lin. Fl. Mansh. (1939) 53; Fl. Kirgiz. 1 (1952) 89; Grubov, Konsp. fl. MNR (1955) 59; Fl. Kazakhst. 1 (1956) 94; Fl. Tadzh. 1 (1957) 96; Chen and Chou, Rast. pokrov r. Sulekhe (1957) 91. —Ic.: Pflanzenr. 31, IV, 11, fig. 27J, K, P; Hagstr. l.c. figs. 54, 55.

Described from Europe. Type in London (Linn.).

Small ponds with stagnant water.

IA. **Mongolia:** *Cis-Hing.* (near Yaksha railway station, meadow lake, Aug. 19, 1902—Litw.), *Cent. Khalkha* (in water of pond in hummocky sand near Borokhchin lake, July 6, 1924—Pavl.), *Bas. Lakes* (Dzabkhyn river near Khirgis-Nur lake, Aug. 2 and 8, 1879—Pot.; on bank of Khorgon-Shibira river, in pond, July 1, 1892—Kryl.), *Val. Lakes* (on right stony bank of Tuin-Gol, opposite Boro-Khoto ruins, in puddle, July 8, 1893—Klem.), *Gobi-Alt.* (Artsa-Bogdo, in small stream, No. 297A, 1925—Chaney), *Khesi* (Sulekhe river—Chen and Chou, l.c.).

IB. **Kashgar:** *Nor.* (Uchturfan, May 14, 1908—Divn.; between Kucha and Kurla, near Bugur town, in solonetz swamps, Aug. 20, 1929—Pop.), *East.* (10 km north-east of Toksun town, on water, No. 7267, June 15, 1958—Lee and Chu).

IIA. **Junggar:** *Cis-Alt.* (Altay [Shara-Sume], 990 m, No. 2351, Aug. 15, 1956—Ching), *Tarb.* (north of Dachen [Chuguchak] town, No. 2959, Aug. 14, 1957—Kuan), *Tien Shan* (B. Yuldus, on swamp near water, 2460 m, No. 6475, Aug. 10, 1958—Lee and Chu), *Jung. Gobi* (Savan district, along course of Manas river, No. 1593, June 29, 1957—Kuan).

IIIB. **Tibet:** *Weitzan* (Huang He river upper course, June 2, 1880—Przew.).

General distribution: Aral-Casp., Fore Balkh., North Tien Shan; Arct. (Europ.), Europe, Mediterranean., Fore Asia, Caucasus, Mid. Asia, West. and East. Sib., North. Mongolia (Hent., Hang., Mong.-Daur.), China (Dunbei), Himalayas (Kashmir), Korean peninsula, Japan, North and South America, Africa (nor. and south.).

2. Ruppia L.
Sp. pl. (1753) 127; Gen. pl. (1737) 277.

1. **R. maritima** L. Sp. pl. (1753) 127; Forbes and Hemsley, Index Fl. Sin. 3 (1903) 197; Aschers. and Graebn. in Engler, Pflanzenr. 31, IV, 11 (1907) 142; Hagstr. in Hedin, S. Tibet, 6, 3 (1922) 97; Krylov, Fl. Zap. Sib. 1 (1927) 114; Juzepczuk in Fl. SSSR, 1 (1934) 262; Kitag. Lin. Fl. Mansh. (1939) 54; Norlindh, Fl. mong. steppe, 1 (1949) 46; Fl. Kazakhst. 1 (1956) 98; Fl. Tadzh. 1 (1957) 100. —Ic.: Pflanzenr. 31, IV, 11, fig. 30.

Described from Europe. Type in London (Linn.).

Along seashores, freshening water bodies near seashores, rarely far from sea, but usually on saline or brackish water lakes.

IA. **Mongolia:** *Alash. Gobi* (Chirgu-Bulyk well, on water, common, Aug. 10, 1873—Przew.; "Gashun-nor, Camp XLIX D, flor. et fruct., Nov. 6, 1927, No. 1730, Hummel"—Norlindh, l.c.).

IIIA. **Qinghai:** *Nanshan* ("Kellajan-ak, Atjet-bulak, salt lake, July 1, 1900, Hedin"—Hagstr. l.c.).

General distribution: subcosmopolitan (temperate and subtropical zones).

3. Zannichellia L.
Sp. pl. (1753) 969; Gen. pl. (1737) 278.

1. **Z. pedunculata** Reichb. in Mössl. Handb. ed. 2, 3 (1829) 1591; Juzepczuk in Fl. SSSR, 1 (1934) 264; Kitag. Lin. Fl. Mansh. (1939) 54; Grubov, Konsp. fl. MNR (1955) 59; Fl. Kazakhst. 1 (1956) 59; Ikonnikov, Opred. rast. Pamira (1963) 41. —*Z. palustris* subsp. *pedicellata* Wahlenb. et Rosen in Nova Acta Upsal, 8 (1821) 227, 254; Hook. f. Fl. Brit. India, 6 (1893) 568; Aschers. and Graebn. in Engler, Pflanzenr. 31, IV, 11 (1907) 156; Hagstr. in Hedin, S. Tibet, 6, 3 (1922) 97. —*Z. palustris* L. s.l.; Persson in Bot. notiser (1938) 273; Norlindh, Fl. mong. steppe, 1 (1949) 47; Fl. Tadzh. 1 (1957) 101.

Described from Europe. Type in Leipzig (?).

Lakes, brooks, ditches, on freshwater and brackish water.

IA. Mongolia: *Cent. Khalkha* (environs of Ikhe-Tukhum-nor lake, Ata-bulak—Mishik-gun, June 1926—Zam,), *East. Gobi* (Khonin-chaghan-chölo-gol, Camp XI, fruct., Aug. 1, 1927, No. 1324, Hummel; Bayan-shandai-sume, Camp XXI, fert. Aug. 16, 1927, No. 1454, Hummel; Khomin-chaghan-chölo-gol, Camp XVI, fert. July 22, 1927, No. 6727, Söder bom'!—Norlindh, l.c.), *Alash. Gobi* (Suvan'no area, on river-overflow lake, July 6, 1886—Pot.; Edzin-Gol river valley, near upper Ontsin-Gol, Bukhan-Khub area, in water-filled pit, July 13, 1926—Glag.).

IB. Kashgar: *West.* ("Kashgar, by riverside, June 13, 1933, No. 477"—Persson, l.c.).

IIIB. Tibet: *Chang Tang* ("Camp 78, on eastern side of Naktsongtso, 4636 m, Sept. 11, 1901, Hedin"—Hangstr. l.c.).

IIIC. Pamir ("Eastern Pamir, upper Basik-kul, freshwater, 3720 m, July 24, 1894, Hedin"—Hagstr. l.c.).

General distribution: Europe, Caucasus, Mid. Asia, West. Siberia (south.), East. Siberia (south.), North Mongolia (Hang.), China (Dunbei, North-west., South-west.), Africa.

Note. This species, like the genus as a whole, has been little studied and calls for special investigation. The distribuiton range of this species is difficult to establish since many investigators do not recognise it and include it under *Z. palustris* L. s.l.

Family 14. ZOSTERACEAE Dum.

1. Zostera L.
Sp. pl. (1753) 968.

1. **Z. marina** L. Sp. pl. (1753) 968; Deasy, in Tibet and Chin. Turk. (1901) 404; Forbes and Hemsley, Index Fl. Sin. 3 (1903) 197; Aschers. and Graebn. in Engler, Pflanzenr. 31, IV, 11 (1907) 28; Juzepczuk in Fl. SSSR, 1 (1934) 266; Kitag. Lin. Fl. Mansh. (1939) 54; Ikonnikov in Novosti sist. vyssh. rast. 1969 (1970) 260.

Described from Baltic Sea. Type in London (Linn.).

On sea coasts, sandy and silty floor up to 10 m depth, sometimes in river deltas.

IIIB. Tibet: *Chang Tang* ("Aksai Chui near Yepal Ungur, 4455 m"—Deasy, l.c.).

General distribution: East. Pamir, Arct., Europe, Balk.-Asia Minor, Caucasus, Far East, China (Dunbei, North), North America, Africa.

Family 15. NAJADACEAE Juss.

1. Najas L.
Sp. pl. (1753) 1015.

1. **N. marina** L. Sp. pl. (1753) 1015; Forbes and Hemsley, Index Fl. Sin. 3 (1903) 198; Krylov, Fl. Zap. Sib. 1 (1927) 116; Juzepczuk in Fl. SSSR, 1 (1934) 270; Kitag. Lin. Fl. Mansh. (1939) 55; Fl. Kirgiz. 1 (1952) 90; Fl. Kazakhst. 1 (1956) 101; Fl. Tadzh. 1 (1957) 102. —*N. major* All. Fl. Pedem. (1785) 221; Hook. f. Fl. Brit. India, 6 (1893) 569. —**Ic.:** Rendle in Trans. Linn. Soc. ser. 2, 5 (1899), pl. 39, fig. 1–30.

Described from Europe. Type in London (Linn.).

In lakes, drowned river valleys and meanders, in fresh or brackish water.

IB. Kashgar: *Nor.* (between Kashgar and Maralbashi, Chonza village, in current through poplar forest, No. 715, Aug. 1; south. Tien Shan foothill, near Ai-Kul' village, in current, Aug. 8—1929, Pop.; in Yuili [Chiglyk] town region, In'so, on water, No. 8625, Aug. 15, 1958—Lee and Chu).

General distribution: Aral-Casp., Fore Balkh., North. Tien Shan (Issyk-Kul'); Europe, Mediterr., Caucasus, Mid. Asia, West. Sibria, East. Siberia (south.), Far East (south.), China (Dunbei, North). Korean peninsula, Japan, Indo-Mal., North America, Africa (nor.), Australia.

Family 16. JUNCAGINACEAE Rich.

1. Triglochin L.
Sp. pl. (1753) 338.

1. Pistils and carpels 6; carpels ovate-prismatic (length not more than twice breadth). Inflorescence compact, with many flowers on relatively short pedicels. Stem compact, about 3 mm thick and up to 20 cm long; leaves fleshy, up to 2 mm broad, and form compact radical cluster. .... 1. **T. maritimum** L.

+ Pistils and carpels 3; carpels clavate or linear-prismatic (length more than 3–5 times breadth). Inflorescence lax with few (up to 10) flowers on slender, obliquely separated pedicels. Stem slender, up to 1 mm thick; leaves thin, up to 1 mm broad, few. ...... 2. **T. palustre** L.

1. **T. maritimum** L. Sp. pl. (1753) 339; Henders. and Hume, Lahore to Jarkand (1873) 337; Franch. Pl. David. 1 (1884) 315; Kanitz in Szechenyi, Wissenschaft. Ergebn. 2 (1898) 735; Hemsley in J. Linn Soc. London (Bot.) 30 (1894) 124; Alcock, Rep. natur. hist. results Pamir Boundary Commiss. (1898) 27; Forbes and Hemsley, Index Fl. Sin. 3 (1903) 192; Buchenau in Engler, Pflanzenr. 16, IV, 14 (1903) 8; Strachey, Catal. (1906) 196; Danguy in Bull. Mus. nat. hist. natur. 17 (1911) 450, 20 (1914) 143; Ostenf. in Hedin, S. Tibet, 6, 3 (1922) 95; Krylov, Fl. Zap. Sib. 1 (1927) 119; Pampanini, Fl. Carac. (1930) 70; Rehder in J. Arn. Arb. 14 (1933) 3; B. Fedtsch. in Fl. SSSR, 1 (1934) 276; Persson in Bot. notiser (1938) 273; Hao in Engler's Bot. Jahrb. 68 (1938)

579; Kitag. Lin. Fl. Mansh. (1939) 55; Walker in Contribs U.S. Nat. Herb. 28 (1941) 595; Norlindh, Fl. mong. steppe, 1 (1949) 48; Fl. Kirgiz. 1 (1952) 91; Grubov, Konsp. fl. MNR (1955) 59; Fl. Kazakhst. 1 (1956) 104; Fl. Tadzh. 1 (1957) 103; Chen and Chou, Rast. pokrov r. Sulekhe (1957) 91; Ikonnikov, Opred. rast. Pamira (1963) 43. —**Ic.:** Fl. Tadzh. 1, Plate XIV, figs. 3, 4.

Described from Europe. Type in London (Linn.).

Solonchak meadows and swamps, along solonchak banks of rivers, lakes and springs.

IA. Mongolia: *Mong. Alt.* (2–3 km south-east of Yusun-Bulak, solonchak-like lowland, July 13, 1947—Tuvanzhab), *Cis-Hing.* (vicinity of Manchuria station, solonetz wet area, No. 926, June 25, 1951—A.R. Lee (1959)), *Cent. Khalkha, East. Mong., Bas. Lakes, Val. Lakes, Gobi-Alt., East. Gobi, West. Gobi* (Tsagan-Bulak area, on south. foothill of Seksek-Bogdo mountain range, solonchak-like low-grass meadow along spring, Aug. 1, 1943—Yun.), *Alash. Gobi, Ordos* (25 km south-east of Otok town, solonchak-like meadow near Khaolaitai lake, Aug. 1; 50 km west-north-west of Hangai town, solonchak meadow on bank of Yan'khaitszy lake, Aug. 6—1957, Petr.), *Khesi.*

IB. Kashgar: *Nor.* (Uchturfan, in garden, May 14, 1908—Divn.; southern base of Tien Shan, near Dzhurga village, in swampy hollows near spring discharges, Aug. 13, 1929—Pop.; Tsinzyaotszin-Pichan, No. 473, May 22, 1957—Kuan; in Aksu region on Muzart river, No. 8288, Sept. 9, 1958—Lee and Chu), *South.* (Keriya river valley 5–6 km from Keriya settlement, along road to Niyu, silted meander, marshy meadow, May 9, 1958—Yun. et al.), *Takla-Makan* (3 km nor. of Cherchen town, around puddle, No. 9521, June 14, 1959—Lee et al.).

IC. Qaidam: *plains.*

IIA. Junggar: *Cis-Alt.* (Koktogoi, No. 1040b, June 7, 1959—A.R. Lee (1959)), *Tien Shan, Jung. Gobi, Dzhark.* (Khoyur-Sumun south of Kul'dzha, May 24; Kul'dzha, July 5—1877), A. Reg.), *Balkh.-Alak.* (Toli district, No. 2498, Aug. 4, 1957—Kuan).

IIIA. Qinghai: *Nanshan* (Nanshan upland, 2400–2550 m, on clayey-solonetz soil, July 10, 1879—Przew.; Humboldt mountain range, Kuku-Usu river, wet meadow, June 6, 1894—Rob.; "La Chi Tzu Shan, No. 695, Ching"—Walker, l.c.). *Amdo* (Baga-gorgi river, 2700 m, May 23; Mudzhik river, 3150–3450 m, on saline soil, June 24—1880, Przew.; "Radja and Yellow River Gorges, rock"—Rehder, l.c.).

IIIB. Tibet: *Chang Tang* ("Karakash Valley, Jarkand, 3900–4800 m"—Henders. and Hume, l.c.; Karakash river upper course, 12 km west of Shakhidull, along road to Kirgiz-Dzhangil pass, 3900 m, short-grass meadow on sandy alluvium, June 3, 1959—Yun. et al.; Sin'tszyan-Tibet highway, 27 km west of Maigaity, No. 483, June 3, 1959—A.R. Lee (1959); "Kuenlun Plains at about 5100 m, Picot"—Hemsley [1894], l.c.), *Weitzan, South.* (Khambajong. Aug. 31, 1903—Younghusband).

IIIC. Pamir (Charlysh river, marsh on river bank, June 21, 1909—Divn.; Sarykol river 25 km south of Bulunkul' along road to Tashkurgan, 3800 m, valley floor, solonchak meadow, June 12, 1959—Yun. et al.; Tashkurgan, in valley, 3000 m, No. 300, June 13, 1959—A.R. Lee (1959); "Pamir region, 3900–4200 m, No. 17763"—Alcock, l.c.; "Pamir, Tashkorgan, Dafdar, ca. 3510 m, June 28, 1935, No. 661"—Persson, l.c.).

General distribution: Aral-Casp., Fore Balkh., Nor. and Cent. Tien Shan, East. Pam.; Arctic, Europe, Mediterranean, Balk.-Asia Minor, Caucasus, Mid. Asia, West. and East. Sib., Far East (south.), Nor. Mongolia (Hent., Hang., Mong.-Daur.), China (Dunbei, North, Northwest), Japan, North and South America.

2. **T. palustre** L. Sp. pl. (1753) 338; Henders. and Hume, Lahore to Jarkand (1873) 337; Kanitz in Szechenyi, Wissenschaft. Ergebn. 2 (1898)

735; Hemsley in J. Linn. Soc. London (Bot.) 30 (1894) 119; Deasy, In Tibet and Chin. Turk. (1901) 404; Hemsley, Fl. Tibet (1902) 200; Forbes and Hemsley, Index Fl. Sin. 3 (1903) 192; Buchenau in Engler, Pflanzenr. 16, IV, 14 (1903) 9; Danguy in Bull. Mus. nat. hist. natur. 6 (1911) 5; Krylov, Fl. Zap. Sib. 1 (1927) 118; Pampanini, Fl. Carac. (1930) 70; B. Fedtsch. in Fl. SSSR, 1 (1934) 277; Persson in Bot. notiser (1938) 273; Kitag. Lin. Fl. Mansh. (1939) 56; Norlindh, Fl. mong. steppe, 1 (1949) 49; Fl. Kirgiz. 1 (1952) 91; Grubov, Konsp. fl. MNR (1955) 60; Fl. Kazakhst. 1 (1956) 104; Chen and Chou, Rast. pokrov r. Sulekhe (1957) 91; Fl. Tadzh. 1 (1957) 104; Ikonnikov, Opred. rast. Pamira (1963) 41. —Ic.: Fl. Tadzh. 1, Plate XIV, figs. 1, 2.

Described from Europe. Type in London (Linn.).

Wet and marshy banks of rivers and lakes on solonchak-like meadows, on coastal pubble beds.

IA. Mongolia: *Mong. Alt., Cis-Hing.* (Yaksha station, on roadside meadow, wet solonetz area, No. 2842, 1954—Wang), *Cent. Khalkha, East. Mong., Bas. Lakes, Val. Lakes, Gobi Alt.* (Leg river, Aug. 28, 1886—Pot., Bain-Tukhum area, hummocky solonchaks north of lake, Aug. 6, 1931—Ik.-Gal.), *East. Gobi, West. Gobi* (Bilgekhu-Bulak area 30–35 km east of Tsagan-Bogdo town, solonchak-like meadow near spring, July 31, 1943—Yun.), *Alash. Gobi, Ordos, Khesi.*

IB. Kashgar: *Nor., West., South.*

IC. Qaidam: *plains, montane.*

IIA. Junggar: *Cis-Alt., Tien Shan, Jung. Gobi, Zaisan* (Ch. Irtysh, right bank below Burchum river, willow scrub, June 15; between Kara area and Kaba village, on bank of irrigation ditch, June 16—1914, Schischk.), *Balkh.-Alak.* (vicinity of Dachen [Chuguchak] town, No. 2850, Aug. 10; same site, No. 2859, Aug. 11—1957, Kuan).

IIIA. Qinghai: *Nanshan* (Kuku-Usu area, on wet meadow, June 6, 1894—Rob.), *Amdo* (upper course of Huang He river, Kha-Gomi area, 2100 m, May 30, 1880—Przew.).

IIIB. Tibet: *Chang Tang* (between Huang He and Yantszy rivers, south. slope, 4200 m, June 12, 1884—Przew.; "Northern Tibet, near Camp 94, 4800 m, July 25, 1898"—Deasy, l.c.), *South.* ("between Tisum and Sutlej river, 4500 m, Strachey and Winterbottom"—Hemsley, l.c.).

IIIC. Pamir (Tagarma valley, Swampy areas, July 23, 1913—Knorr.; 3–4 km west of Tashkurgan, meadow gullies, June 13; 100 km south of Kashgar along road to Tashkurgan, hot spring, June 15—1959, Yun.).

General distribution: Aral-Casp., Fore Balkh., Jung.-Tarb., Nor. and Cent. Tien Shan, East. Pamir; Arctic, Europe, Caucasus, Mid. Asia, West. and East. Sib., Far East, Nor. Mongolia, China (Dunbei, North, North-west, South-west), Himalayas, Korean peninsula, Japan, North America, South America (Chile).

Family 17. ALISMATACEAE Vent.

1. Inflorescence whorled (3 in verticel); flowers unisexual; stamens several. Receptacle convex. ....2. **Sagittaria** L.
+   Inflorescence large, paniculate; flowers bisexual; stamens 6. Receptacle flat. .... 1. **Alisma** L.

1. Alisma L.

Sp. pl. (1753) 342

1. Leaves lanceolate to ovate-lanceolate and broadly ovate, rounded or weakly cordate at base. Furrow on mature fruits not deep (0.1 mm or less); fruits usually with 1 furrow ....
.... 3. **A. plantago-aquatica** L.
+ Leaves ribbon-like, linear, narrowly lanceolate or lanceolate, gradually narrowing clavately at base into petiole .... 2.
2. Leaves lanceolate with enlarged base and strongly attenuated tip. Fruits with 1 or 2 furrows .... 2. **A. lanceolatum** With.
+ Leaves ribbon-like, linear or linear-lanceolate. Fruits invariably with 2 furrows. Inflorescence lax, spreading. ..... 1. **A. gramineum** Gmel.

1. **A. gramineum** Gmel. Fl. Badensis Alsatica, 4 (1826) 256; Samuelss. in Arkiv. Bot. 24, 7 (1932) 35; id. in Pflanzenareale, 3, 8 (1933) Karte 75; Kitag. Lin. Fl. Mansh. (1939) 56; Norlindh, Fl. mong. steppe, 1 (1949) 51; Hendricks in Amer. Midl. Natur. 58, 2 (1957) 485. —*A. loeselii* Gorski in Eichw. Nat. Skizze Lith. 1 (1830) 127, nomen; Juzepczuk in Fl. SSSR, 1 (1934) 282; Grubov, Konsp. fl. MNR (1955) 60; Fl. Kazakhst. 1 (1956) 106; Fl. Tadzh. 1 (1957) 108. —*A. arcuatum* Michal. in Bull. Soc. bot. France, 1 (1854) 312; Krylov, Fl. Zap. Sib. 1 (1927) 122. —*A. plantago* β *angustifolia* auct. non Ledeb.: Danguy in Bull. Mus. nat. hist. natur. 17 (1911) 450. —**Ic.:** Samuelss. l.c. (1932) tabs. 4, 5; Hendricks, l.c. figs. 9–11.

Described from Europe (from Elzas-Baden region). Type probably lost.

Swamps, wet meadows, river banks as well as lakes, ponds, irrigation ditches; common on saline soils.

IA. Mongolia: *East. Mong.* (Mongolia chinensis, 1831—l. Kuznetsov; Sartchy [Salachi], June 1866—David; vicinity of Sinbaerkhuyunchi town, on water, No. 1062, June 30; vicinity of Khailar town, on shallow water, No. 1129, July 4—1951, S.H. Li et al. (1951); "Ad rivum Djaghastei, in aqua, Aug. 13, 1920, No. 837, Andersson"—Norlindh, l.c.), *Bas. Lakes* (near Khirgis-Nor lake, Dzabkhyn river Aug. 8, 1879—Pot.; Shargain-Gobi, pass from Borin-Khotok collective to Gol-Ikhe, stone trail, Sept. 4; Shargain-Gol bank between Shargain-Nor lake and Tszak-Obo, solonchaks, Sept. 8—1930, Pob.), *Alash. Gobi* (Kheikho river, June 25; Khara-Sukhai river, July 20—1886, Pot.; "Bayan-Bogdo Camp, Wentsunhai-tze, Sept. 1928, No. 7187; eod. loco. June 22, 1929, No. 7628, Söderbom"—Norlindh, l.c.).

IB. Kashgar: *Nor.* (near Bugur village, in marshy areas, Aug. 20, 1929—Pop.; in Yuili [Chiglyk] region, near lake, No. 8599, Aug. 12, 1958—Lee and Chu).

IIA. Junggar: *Jung. Gobi* (Urungu river, Aug. 21, 1876—Pot.; Savan district, Shamyntsza, No. 948, June 17, 1957—Kuan), *Zaisan* (right bank of Ch. Irtysh below Burchum river, Sary-Dzhasyk-kol, Kiikpai, willow scrub, June 15, 1914—Schischk.).

IIIA. Qinghai: *Nanshan* ("Jong-Ngan, marais, alt. 3200 m, route de Kantcheou a Si-Ning, July 3, 1908, Vaillant"—Danguy, l.c.).

General distribution: Aral-Casp., Fore Balkh, Jung-Tarb.; Europe, Mediterranean, Balk.-Asia Minor, Caucasus, Mid.Asia, West. Sib. (south.), East. Sib. (south.), China (Dunbei), North America.

Note. This species grows very well in water as well as on land and is highly variable. Many varieties, rather common among amphibious species, are known. They differ in general habit as well as in shape of leaves (oblong-elliptical, lanceolate, linear or ribbon-like) and length of petiole (petiole shorter, as long as or much longer than leaf blade) depending on where the plant grew; its leaves float or submerge; submerged leaves lacking in plants growing in shallow water.

2. **A. lanceolatum** With. Bot. Arang. Brit. Plants, ed. 3, 2 (1796) 262; Samuelss. in Arkiv. Bot. 24, 7 (1932) 21; id. in Pflanzenareale, 3, 8 (1933) Karte 74; Juzepczuk in Fl. SSSR, 1 (1934) 281; Fl. Kirgiz. 1 (1952) 95; Fl. Kazakhst. 1 (1956) 106; Fl. Tadzh. 1 (1957) 107; Hendricks in Amer. Midl. Natur. 58, 2 (1957) 490. —**Ic.:** Samuelss. l.c. (1932) tab. 1; Hendricks, l.c. fig. 12.

Described from Europe (British Isles). Type in London (BM).

On wet banks of rivers, lakes and ponds.

IIA. Junggar: *Jung. Gobi* (3 km east of Kuitun town, No. 424, Kuan—doubtful specimen).

General distribution: Aral-Casp., Fore Balkh.; Europe, Mediterranean, Balk.-Asia Minor, Fore Asia, Caucasus, Mid. Asia, Himalayas (Kashmir).

3. **A. plantago-aquatica** L. Sp. pl. (1753) 342; Franch. Pl. David. 1 (1884) 314; Hook. f. Fl. Brit. India, 6 (1893) 559; Forbes and Hemsley, Index Fl. Sin. 3 (1903) 189; Danguy in Bull. Mus. nat. hist. natur. 20 (1914) 143; Krylov, Fl. Zap. Sib. 1 (1927) 121; Pampanini, Fl. Carac. (1930) 70; Samuelss. in Arkiv Bot. 24, 7 (1932) 12; id. in Pflanzenareale, 3, 8 (1933) Karte 73; Juzepczuk in Fl. SSSR, 1 (1934) 280; Hao in Engler's Bot. Jahrb. 68 (1938) 579; Fl. Kirgiz. 1 (1952) 95; Grubov, Konsp. fl. MNR (1955) 60; Fl. Kazakhst. 1 (1956) 105; Fl. Tadzh. 1 (1957) 107; Hendricks in Amer. Midl. natur. 58, 2 (1957) 475; Chen and Chou, Rast. pokrov r. Sulekhe (1957) 91. —*A. plantago-aquatica* var. *orientale* Samuelss. in Acta Horti Gothob. 2 (1926) 84. —*A. plantago-aquatica* ssp. *orientale* Samuelss. in Arkiv Bot. 24, 7 (1932) 16; Norlindh, Fl. mong. steppe, 1 (1949) 50. —*A. orientale* (Samuelss.) Juz. in Fl. SSSR, 1 (1934) 281; Kitag. Lin. Fl. Mansh. (1939) 56; Fl. Kirgiz. 1 (1952) 95. —**Ic.:** Hendricks, l.c. figs. 5–7.

Described from Europe. Type in London (Linn.).

Marshy banks of rivers and lakes, in ditches, swamps.

IA. Mongolia: *East. Mong.* (Mongolia chinensis, 1831—I. Kuznetsov; Huang He river before Khekou, Aug. 4, 1884—Pot.; Khailar town, brackish water reservoir, No. 2978, 1954—Wang; "Doyen, in aqua stagnante, Aug. 5, 1926, Eriksson"—Norlindh, l.c.), *Bas. Lakes* (near Ubsa lake, Ulangom monastery, Sept. 6, 1879—Pot.), *Alash. Gobi* (south. Alashan, Bayan-Bulyk cliff, Aug. 29, 1880—Przew.), *Ordos* (Huang He river valley, Aug. 6, 1871—Przew.; 10 km south-west of Ushin town, on bank of lake, among scrub, Aug. 4, 1957—Petr.), *Khesi* (Suchzhou [Tszyutsyuan'], June 25, 1890—Marten; Tszyutsyuan', Baidekhe, No. 42, July 2, 1956—Ching; Sulekhe river—Chen and Chou, l.c.).

IB. Kashgar: *Nor.* (south of Yuili town, in water, No. 8541, Aug. 5, 1958—Lee and Chu (A.R. Lee (1959)), *East.* (Khami oasis, June 13, 1879—Przew.; Bugas area, 480 m, Aug. 23, 1895—Rob.).

IIA. Junggar: *Cis-Alt.* (Quinhe [Chingil'], No. 1635, Aug. 11; Barbagai-Burchum, No. 2888, Sept. 11—1956, Ching), *Jung. Alt.* upper course of Borotal, 1500–1800 m, Aug. 4, 1879—

A. Reg.), *Tien Shan* (Tekes river, near river on wet clay, July 10, 1893—Rob.; Sin'yuan', No. 3758, Aug. 22, 1957—Kuan), *Jung. Gobi* (Ch. Irtysh, Aug. 26, 1876—Pot.; Tsitai district, Nan'khu lake, No. 660, July 24, 1956—Ching; Savan district, Mogukhu water reservoir, No. 1559, June 25; near Kuitun town, on farm hedge, No. 267, June 29; 2 km north of Kuitun town, No. 403, July 6; 2 km east of Kuitun town, No. 426, July 7—1957, Kuan), *Zaisan* ("bords de l'Irtich, Aug. 29, 1895, Chaffanjon"—Danguy, l.c.), *Dzhark.* (Kul'dzha, May 27, 1877; Ili river, Kul'dzha, June 1878—A. Reg.; near Suidun, July 1886—Krasnov).

IIIA. Qinghai: *Nanshan* (Lovachen town, in swamp on silty soil, July 26, 1909—Czet.; "Lo-tu-hsien bei Kao-miao, 2000 m, im Wasser oder am Wasser, 1930, Hao"—Hao, l.c.).

IIIB. Tibet: *Weitzan* ("Amne Matchin, im Süsswasser, 4500 m, 1930"—Hao, l.c.).

General distribution: Aral-Casp., Fore Balkh., Jung.-Tarb., Nor. and Cent. Tien Shan; Europe, Mediterranean, Balk.-Asia Minor, Caucasus, Mid. Asia, West. and East. Sib.; Far East, Nor. Mongolia (Hang.), China (Dunbei, North, North-west, Central, South-west), Himalayas (west.), Korean peninsula, Japan, North and South America, Africa, Australia.

Note. Samuelsson (1926, l.c.) described variety *A. plantago-aquatica* var. *orientale* from Yunnan. He later (1932) raised it to subspecies rank while S.V. Juzepczuk (1934, l.c.) elevated it to rank of species—*A. orientale* (Samuelss.) Juz.—with the following distribution range: Far East (Ussur.), Japan, Mongolia, China and the Himalayas.

After studying a large amount of material for species *A. plantago-aquatica* as a whole and species *A. orientale*, we concluded that *A. orientale* should more correctly be considered a variety of the former. Typical specimens of *A. orientale* are mainly confined to the south-eastern part of the distribution range of *A. plantago-aquatica* and found there along with it. Var. *orientale* has no distinct distribution range as such but is scattered over much of the range of *A. plantago-aquatica*.

2. Sagittaria L.
Sp. pl. (1753) 993.

1. **S. trifolia** L. Sp. pl. (1753) 993; Gorodkov in Tr. Bot. muzeya Akad. nauk, 10 (1913) 156; Krylov, Fl. Zap. Sib. 1 (1927) 125; Juzepczuk in Fl. SSSR, 1 (1934) 288; Kitag. Lin. Fl. Mansh. (1939) 56; Fl. Kirgiz. 1 (1952) 92; Fl. Kazakhst. 1 (1956) 108; Fl. Tadzh. 1 (1957) 110. —*S. sagittifolia* auct. non L.: Hook. f. Fl. Brit. India, 6 (1893) 561 p.p.; Forbes and Hemsley, Index Fl. Sin. 3 (1903) 190 p.p. —*S. sagittifolia* β *longiloba* Turcz. Fl. baic.-dah. 2 (1856) 153; Danguy in Bull. Mus. nat. hist. natur. 17 (1911) 5 and 20 (1914) 143. —**Ic.:** Fl. SSSR, 1, Plate XIV, fig. 8.

Described from China based on Petiver's diagram [Petiver, Gazophylac. Nat. et Artis (1702) t. 19, f. 5].

Along banks of rivers, lakes, irrigation ditches as well as in water of closed or poorly running water reservoirs.

IA. Mongolia: *East. Mong.* (environs of Kuku-Khoto [Khukh-Khoto] July 18; Huang He river before Khekou—1884, Pot.; vicinity of Khailar town, brackish water reservoir, No. 2994, 1954—Wang), *Ordos* (Huang He valley, Aug. 1, 1884—Przew.), *Khesi* (Kheikho river near Gaotai, June 20, 1886—Pot.; Soutcheou, marais, alt. 1600 m, June 23, 1908, Vaillant"—Danguy [1911], l.c.).

IB. Kashgar: *Nor.* (4 km from Yuili [Chiglyk], in irrigation ditch, No. 8518, Aug. 2, 1958—Lee and Chu (A.R. Lee (1959)), *East* (Kuruktag mountain range, on Kurle river near Bash-Akin Picket, Aug. 24, 1929—Pop.).

IIA. Junggar: *Jung. Gobi* (Barbagai-Burchum, No. 2887, Sept. 11, 1956—Ching), *Zaisan* ("Bords de l'Irtich, No. 1170, Aug. 20, 1895, Chaffanjon"—Danguy [1914], l.c.), *Balkh.-Alak.* (Emel' river valley, meanders, June 20, 1905—Orbuchev).

General distribution: Aral-Casp., Fore Balkh.; Fore Asia, Caucasus, Mid. Asia, East. Sib. (south. Transbaikal), Far East (south.), China (Dunbei, North, Central, South-west, South, Hainan, Taiwan), Himalayas, Korean peninsula, Japan, Indo-Mal.

Note. The species varies greatly in shape of leaves and their lobes as well as leaf size, but the proportion of leaf sections (blades and lobes) is fairly constant.

Family 18. **BUTOMACEAE** Rich.

1. **Butomus** L.
Sp. pl. (1753) 372.

1. **B. umbellatus** L. Sp. pl. (1753) 372; Franch. Pl. David. 1 (1884) 315; Hook. f. Fl. Brit. India, 6 (1893) 562; Forbes and Hemsley, Index Fl. Sin. 3 (1903) 191; Danguy in Bull. Mus. nat. hist. natur. 20 (1914) 143; Krylov, Fl. Zap. Sib. 1 (1927) 127; B. Fedtsch. in Fl. SSSR, 1 (1934) 292; Kitag. Lin. Fl. Mansh. (1939) 57; Fl. Kirgiz. 1 (1952) 96; Grubov, Konsp. fl. MNR (1955) 60; Fl. Kazakhst. 1 (1956) 109; Fl. Tadzh. 1 (1957) 112. —**Ic.:** Fl. Tadzh. 1, Plate 17.

Described from Europe. Type in London (Linn.).

On shallow lakes, ponds, rivers, meanders, in stagnant and slow-moving waters.

IA. Mongolia: *East. Mong.* (Mongolia chinensis, in swamp, 1831—I. Kuznetsov; Ul'gen-Gol [Uligen, Ulugui] river, on flat bank, July 18, 1899—Pot. and Sold; vicinity of Khailar town, near water, No. 1121, July 4, 1951—S.H. Lee (1951)), *Ordos* (Huang He river valley, Aug. 3, 1871—Przew.).

IIA. Junggar: *Cis-Alt.* (west of Fuyun' [Koktogoi] town, No. 1885, Aug. 11; Qinhe [Chingil'] district, No. 2909, Aug. 13—1956, Ching), *Jung. Gobi* (Savan district, Mogukhu water reservoir, No. 1558, June 25, 1957—Kuan), *Zaisan* (right bank of Ch. Irtysh below Burchum river, Sary-dzhasyk-kol, Kiikpai, willow scrub, June 15, 1914—Schischk.; "Bords de l'Irtich, No. 1176, Aug. 30, 1895, Chaffanjon"—Danguy, l.c.), *Dzhark.* (Ili river in Kul'dzha region, June 1878—Mate; near Suidun, 1886—Krasnov), *Balkh.-Alak.* (Emel' river valley, meanders, June 20, 1905—Obruchev).

General distribution: Aral-Casp., Fore Balkh., Jung.-Tarb., Nor. and Cent. Tien Shan; Arctic (Europ.), Europe, Balk.-Asia Minor, Fore Asia, Caucasus, Mid. Asia, West. Siberia, East. Siberia (west. and south.), Far East (south.), Nor. Mongolia, China (Dunbei, North), Himalayas.

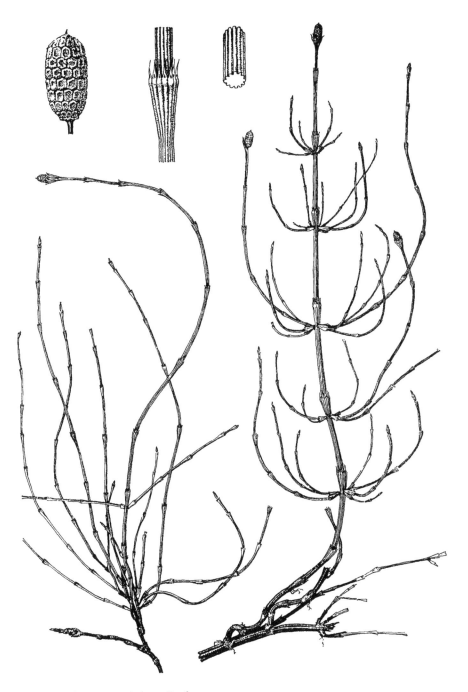

Plate I: *Equisetum ramosissimum* Desf.

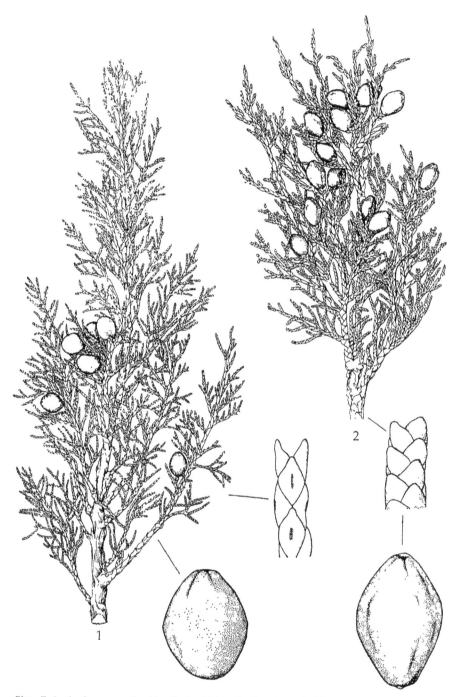

Plate II: 1—*Juniperus pseudosabina* Fisch. et Mey.; 2—*J. turkestanica* Kom.

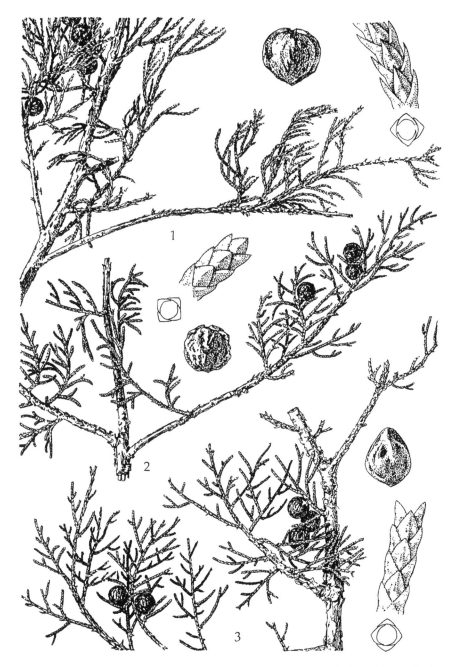

Plate III: 1—*Juniperus przewalskii* Kom.; 2—*J. przewalskii* var. *zaidamensis* (Kom.) Matz.; 3—*J. tibetica* Kom.

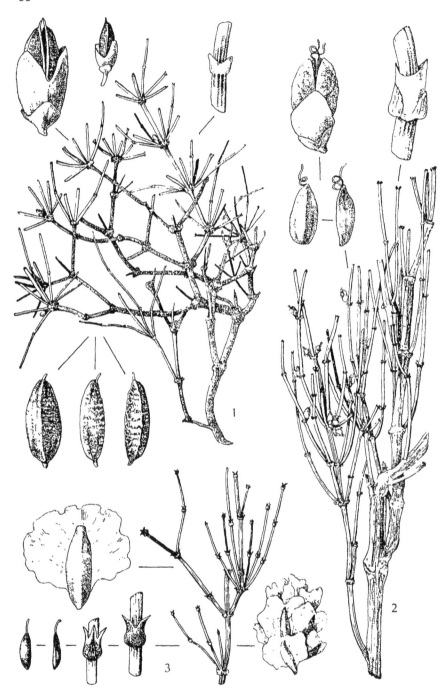

Plate IV: 1—*Ephedra rhytidosperma* Pachom.; 2—*E. glauca* Regel; 3—*E. przewalskii* Stapf.

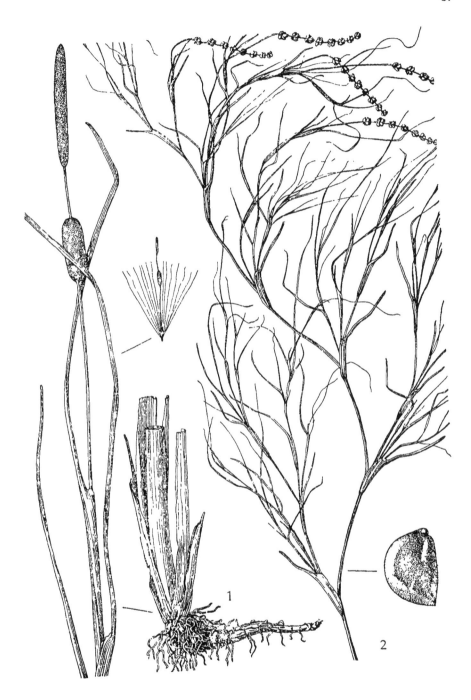

Plate V: 1—*Typha laxmannii* Lepech.; 2—*Potamogeton vaginatus* Turcz.

68

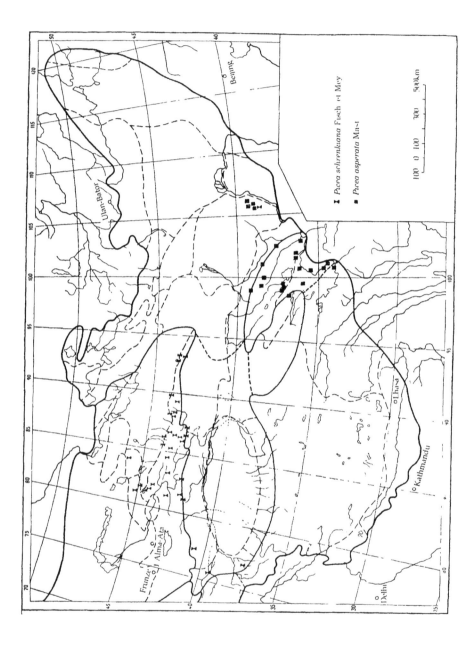

Map 1.

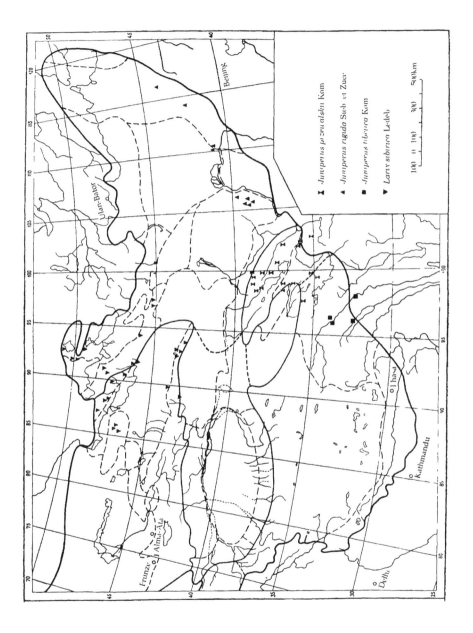

Map 2.

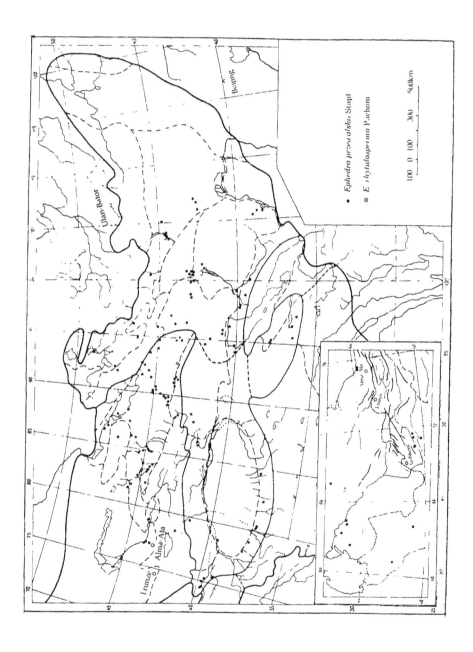

Map 3.

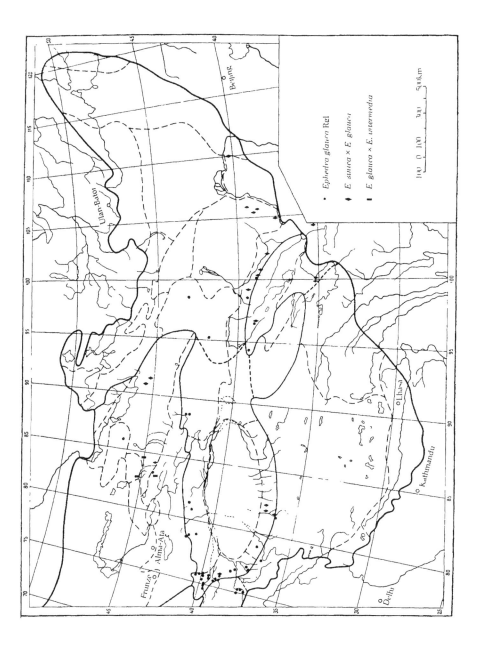

Map 4.

ADDENDA[1]

Important new geographical record
Pinus tabulaeformis Carr.

IA. **Mongolia:** *Alash. Gobi* ("Xiao Tengger Sands" [in Acta Sci. Natur. Univ. Intramongol. 13, 1 (1982) 121–124]).

---

[1]Compiled by V.I. Grubov in 2001

BIBLIOGRAPHY FOR FLORA OF CENTRAL ASIA[1]

Supplement 1

This supplement lists all the works reviewed by us that were published after publication of the first volume of Flora of Central Asia (Russian edition, 1963; English edition, 1999) which contains the main bibliographic references. The supplement also covers some pre-1963 works that were accidentally oversighted. Further, more complete descriptions of several multivolume 'Flora' are given here; they were only mentioned in volume 1 because still incomplete as of 1963.

Ali, S.I. 1963. A taxonomic revision of genus *Gueldenstaedtia* Fisch. Candollea, 18: 137–159. Map.

Banerjee, S.P. 1966. A taxonomic revision of Indian *Trigonotis* Stev. (Boraginaceae). Bull. Bot. Surv. India, 8 (3–4): 319–327.

Banzragch, D. 1964. The study of vegetation in montane-forest-steppe zone (Mongolia). BNMAU Shinzhlekh Ukhaany Akad. medee [Izv. AN MNR], 2: 69–80 (in Mong., Russian summary).

———— 1965. New reports for Mongolian flora. BNMAU Shinzhlekh Ukhaany Akad. medee, 1: 92–93 (in Mong.).

———— 1966. Festuca ovina steppe. Tr. Biol. inst. MNR, 1: 49–61.

Banzragch, D. and Yunatov, A.A. 1967. Some additions to the flora of Northern Hangay. BNMAU Shinzhlekh Ukhaany Akad. medee [Izv. AN MNR], 3: 38–42 (in Mong., Russian summary). See Yunatov and Banzragch.

Baranov, A.I. 1969. The species of *Corispermum* (Chenopodiaceae) in north-eastern China (2). J. Jap. Bot. 44 (7): 195–206.

Barthel, H., Haase, G. and Richter, H. 1962. Die Mongolische Volksrepublik. Zeitschr. Erdkundeunterricht, 14 (8–9): 288–318, 351–359. Map.

Baum, B. 1966. Monographic revision of Genus *Tamarix*. Jerusalem: I–III, 1–193.

Beletsky, E.A. *1963. Through Tibet to Jomolungma foothills. Izv. Vsesoyuzn. geogr. obshch. 95(3): 203–212. Map.

Bobrov, E.G. *1950. Botanical results of researches of Russian scientists in Central Asia. Bot. zh. 35 (4): 432–437.

———— *1965. Origin of Old World desert flora in the context of a review of genus *Nitraria* L. Bot. zh. 50 (8): 1053–1067 (English summary).

---

[1] Compiled by V.I. Grubov.
*Asterisked works are in Russian—General Editor.

———— *1966. Review of genus *Reaumuria* L. in the context of the origin of Afro-Asian desert flora. Bot. zh. 51 (8): 1057–1072. Maps.

———— *1967. Review of genus *Myricaria* Desv. and its history. Bot. zh. 52 (7): 924–936. Maps.

———— *1969. Genus *Gymnocarpos* Forsk. (Paronychioideae) and its species. Bot. zh. 54 (10): 1576–1583. Map.

Böhme, H. 1963. Allgemeiner Bericht über die Mongolisch-Deutsche Biologische Expedition 1962. Kulturpflanze, 11: 26–33.

Boissieu, M.H., de. 1910. Les Ombelliferes de la mission Pelliot-Vaillant. Bull. Mus. nat. hist. natur. 3: 162–166.

Bor, N.L. 1965. New species in Gramineae. Bull. Bot. Surv. India, 7 (1–4): 132–133.

Borissova, A. [Borisova, A.G.] *1964. Notes on genus *Chesneya* Lindl. and new genus *Chesniella* Boriss. (family Leguminosae). Novosti sist. vyssh. rast. 1964: 178–190.

Botschantzev, V. [Bochantsev, V.P.] *1956. The new genera of family Chenopodiaceae. Commemoration of 75th birthday of Acad. V.N. Sukaczev. Sb. rabot po geobot., lesoved., paleogeogr. i florist.: 108–118.

———— *1965. Critical notes on Chenopodiaceae, 2. Novosti sist. vyssh. rast. 1965: 111–113.

———— *1966. Critical notes on Cruciferae, 5. Novosti sist. vyssh. rast. 1966: 122–139.

———— *1968a. Critical notes on Cruciferae, 6. Novosti sist. vyssh. rast. 1968: 140–148.

———— *1968b. One more report of northern desert member in flora of southern Middle Asia. Bot. zh. 53 (1): 94–95.

———— *1969. Genus *Salsola* L., brief history of its development and dispersal. Bot. zh. 54 (7): 989–1001.

Brown, D.F.M. 1964. A monographic study of the fern genus *Woodsia*. Beih. Nova Hedwigia, 16: 1–X, 1–154.

Bykov, B.A. *1962a. Some observations on sparse forests in Ili river valley. Byull. Mosk. obshch. ispyt. prir., otd. biol. 67 (4): 101–108 (English summary).

———— *1962b. Composition of some formations and ingregations. Tr. Inst. bot. AN KazSSR, 13: 3–27. Map.

———— *1968. Origin of sandy vegetation of Middle Asia and Kazakhstan. Problems of Desert Development (Ashkhabad), 1: 12–21 (English summary).

Chang, C.W. 1963. *Potentilla anserina* on the Chinghai plateau. Shengwuxue Tongbao, 1: 56.

Chaudhri, M.N. 1968. A revision of Paronychiinae. Meded. Bot. Bus. Herb. Rijksuniv. Utrecht, 285: 1–440.

Chou Tin-su [Chzhou Tin-zhu] 1963. Main types of Quaternary continental formations in Sinkiang territory and their relation to the development of topography and climate. Acta geogr. sinica, 29 (2): 109–129 (Chin.; Russian summary).

Czukavina, A. [Chukavina, A.P.] *1966. Two new Asian species of genus *Polygonum* L. and forgotten species *Polygonum rottboellioides* Jaub. et Spach. Novosti sist. vyssh. rast. 1966: 87–91.

Danert, S. 1965. Bericht über die 2. Mongolisch-Deutsche biologische Expedition (1964). Kulturpflanze, 13: 28–44. Map.

Dashnyam, B. *1965. Additions to the flora of Eastern Mongolia. Bot. zh. 50 (11): 1638–1642. Map.

Davazamč [Davazhamts], Ts. *1966. Main edible plants of natural pastures and meadows in the northern part of Uburkhangaisk ajmaq. Tr. Biol. inst. AN MNR, 1: 47–75.

Davazamč [Davazhamts], Ts. and Dashnyam, B. 1968. Commemorating Aleksandr Afanas'evich Yunatov (1909–1967). BNMAU Shinzhlekh Ukhaany Akad. medee [Izv. AN MNR], 3: 100–103 (in Mong.).

Dorofeev, P.I. *1965. Some problems of the history of floras. Bot. zh. 50 (11): 1511–1522.

Efremov, Yu.K. *1960. N.K. Rerikh [Roerich] and geography (85th birthday commemoration). Vopr. geogr. 50: 253–256. Map.

Egorova, T. [Egorova, T.V.] *1964. Critical notes on sedges of section Capillares. Novosti sist. vyssh. rast. 1964: 31–48.

——— *1965. Classification of sedges of subgenus *Vignea* (Beauv.) Kirschl. in the flora of USSR. Novosti sist. vyssh. rast. 1965: 57–83.

——— *1966. New species of genus *Carex* L. from Tibet. Novosti sist. vyssh. rast. 1966: 34–35.

Erdenezhav, G. *1963. Brief note on vegetation of Trans-Altay Gobi. Tr. Inst. estestv. nauk, 1: 86–97.

Fedorov, An. [Fedorov, An. A.] *1966. Some species of subgenus *Aleuritia* (Duby) Wendelbo of genus *Primula* L. Novosti sist. vyssh. rast. 1966: 191–196.

Fedtschenko [Fedchenko], B.A. 1916. Verzeichnis der Pflanzen vom Bogdo-Ola Gebierge, gesammelt von Professor Dr. G. Merzbacher. In: Merzbacher G. Die Gebirgsgruppe Bogdo-Ola im ostlichen Tian-Schan. Abh. Bayer. Ak. Wissensch. math.-phys. Kl. 27 (5): 308–312.

Fischer, F.E. and Meyer, C.A. 1841–1842. Enumeratio plantarum novarum a clarissimo Schrenk lectarum. Petropoli. 1: I–VII, 1–113; 2: I–III, 1–77.

Filatova, N.S. *1963. Key to wormwoods of Kazakhstan. Tr. Inst. bot. AN KazSSR, 15: 204–236.

*Flora of Kazakhstan. 1956–1966. Vols. 1–9. Izd. AN KazSSR. Inst. botaniki, Alma-Ata. 1 [Polypodiaceae—Gramineae], (1956): 1–XVIII, 1–354. Map; 2 [Cyperaceae—Orchidaceae. Ispr. i dopoln. k t. 1 (Corrections and Supplement to Vol. 1)], (1958): I–XIV, 1–292; 3 (Salicaceae—Caryophyllaceae. Addenda. Ispr. i dopoln. k tt. 1, 2 (Corrections and Supplement to Vols. 1 and 2)], (1960): I–XVIII, 1–460; 4 [Nymphaeaceae—Rosaceae. Addenda], (1961): I–XXI, 3–548; 5 [Leguminosae, Addenda. Dopoln. i ispr. k tt. 1–5 (Additions and Corrections to Vols. 1–5)], (1961): I–XIII, 1–155; 6 [Geraniaceae—Umbelliferae. Addenda. Dopoln. i ispr. k tt. 3–6 (Additions and Corrections to Vols. 3–6)], (1963), I–XX, 1–465; 7 [Pyrolaceae—Labiatae. Addenda. Dopoln. k t. 6 (Additions to Vol. 6), (1964): I–XXII, 5–497. Map; 8 [Solanaceae—Compositae: Eupatorieae—Helenieae. Addenda], (1965): I–XVI, 1–447; 9 [Compositae: Anthemideae—Cichorieae. Addenda. Dopoln. k t. 6 (Additions to Vol. 6)], (1966): I–XX, 1–640.

*Flora of Kirgiz SSR. Key to Plants of Kirgiz SSR. 1950–1967. Vols. I–XI, Supplement, No. 1. Izd. AN KirgizSSR, Frunze. 1 [Polypodiaceae—Hydrocharitaceae], (1952): 1–104; 2 [Gramineae—Cyperaceae], (1950): 1–315; 3 [Araceae—Orchidaceae], (1951): 1–150; 4 [Salicaceae—Polygonaceae], (1953): 1–156; 5 [Chenopodiaceae—Caryophyllaceae], (1955): 1–186; 6 [Ceratophyllaceae—Cruciferae], (1955): I–IV, 5–299; 7 [Crassulaceae—Cynomoriaceae. Dopoln. k tt. 7, 6 (Supplement to Vols. 7, 6)], (1957): I–IV, 5–643; 8 [Umbelliferae—Convolvulaceae], (1956): 1–223; 9 [Labiatae-Solanaceae. Annex], (1960): 1–214; 10 [Cuscutaceae—Lobeliaceae. Dopoln. k tt. 1, 3, 7, 9 i 10 (Additions to Vols. 1, 3, 7, 9 and 10)], 1962: 1–388; 11 [Compositae. Dopoln. k t. 10. Alfavitnyi ukazatel' k tt. 1–11 (Supplement to Vol. 10. Alphabetical Index to Vols. 1–11), (1965): 1–607; Supplement, No. 1 (1967): 1–149.

*Flora of Transbaikal area. 1929–1954. Nos. 1–6. 1 (1929), Polypodiaceae—Gramineae, Izd. Troitskosavsk. otd. Gos. geogr. obshch., Leningrad: 1–104, Map; 2 (1931), Cyperaceae—Orchidaceae, Izd. Troitskosavsk. otd. Obshch. nauch. Sibiri i Krae. muzeya, Leningrad: 105–168; 3 (1937), Salicaceae—Chenopodiaceae, Izd. AN SSSR, Moscow-Leningrad: 169–288; 4 (1941), Portulacaceae—Papaveraceae, Izd. AN SSSR, Moscow-Leningrad: 289–416; 5 (1949), Cruciferae—Rosaceae, Izd. AN SSSR, Leningrad: 417–540; 6 (1954), Leguminosae, Izd. AN SSSR, Moscow-Leningrad: 541–664.

Flora of the USSR. 1934–1964. Vols. 1–30. Ukazatel' k tt. 1–30 (Index to Vols. 1–30). Izd. AN SSSR, Bot. inst. im. V.L. Komarova, Moscow—Leningrad. 1 (Hymenophyllaceae—Hydrocharitaceae], 1934: I–XVI, 1–302, 2 Maps; 2 [Gramineae], 1934: I–XXXIII, 1–778; 3 [Cyperaceae—Juncaceae], 1935: I–XXV, 1–636; 4 [Liliaceae—Orchidaceae], 1935: I–XXX, 1–760; Alfavitnyi ukazatel' nazvanii rastenii tt. 1–4 (Alphabetical Index of Plant Names for Vols. 1–4) (1936): I–IV, 1–128; 5 [Saururaceae—Polygonaceae], 1936: I–XXVI, 1–762; 6

[Chenopodiaceae—Caryophyllaceae], 1936: I–XXXVI, 1–956; 7 [Nymphaeaceae–Papaveraceae], 1937: I–XXVI, 1–972; 8 [Capparidaceae—Resedaceae], 1939: I–XXX, 1–696; 9 [Droseraceae—Rosaceae: Spiraeoideae—Pomoideae], 1939: I–XIX, 1–524, 2 Maps; 10 [Rosaceae: Rosoideae—Prunoideae], 1941: I–XXIII, 1–675; 11 [Leguminosae: Mimosoideae—Papilionatae], 1941: I–XVII, 1–432, 2 Maps; 12 [Leguminosae: Papilionnatae—Astragalus], 1946: I–XXVIII, 1–919; 13 [Leguminosae: Papilionatae—Oxytropis—Vigna], 1948: I–XXIV, 1–588; 14 [Geraniaceae—Vitaceae], 1949: I–XXIV, 1–792; 15 [Tiliaceae—Cynomoriaceae], 1949: I–XXIII, 1–743; 16 [Araliaceae—Umbelliferae: Hydrocotyle—Cenolophium], 1950: I–XXVI, 1–648; 17 [Umbelliferae—Cornaceae], 1951: I–XIX, 1–392; 18 [Pyrolaceae—Asclepiadaceae], 1952: I–XXIX, 1–803, 2 Maps; 19 [Convolvulaceae—Verbenaceae], 1953: I–XXIV, 1–752; 20 [Labiatae: Ajuga—Neustruevia], 1954: I–XVII, 1–556; 21 [Labiatae: Eremostachys—Ocimum], 1954: I–XXII, 1–704; 22 [Solanaceae—Scrophulariaceae], 1955: I–XXV, 1-861; 23 [Bignoniaceae—Valerianaceae], 1958: I–XXIII, 1–776; 24 [Morinaceae—Lobeliaceae], 1957: I–XVII, 1–502; 25 [Compositae; Eupatorieae—Heliantheae, 1959; X–XXI, 1–630; 26 [Compositae: Anthemideae—Arctotideae], 1961: I–XXIV, 1939; 27 [Compositae: Echinopsideae—Cynareae], 1962: I–XXII, 1–758; 28 [Compositae: Cynareae—Mutisieae], 1963: I–XX, 1–654; 29 [Compositae: Cychorieae], 1964: I–XXIV, 1–798; 30 [Compositae: Hieracium], 1960: I–XXIV, 1–732; Alfavitnyi ukazatel' k tt. 1–30 (Alphabetical Index to Vols. 1–30), 1964: 1–264, 2 Maps.

*Flora of Western Siberia. 1927–1949, 1961, 1964. Vols. 1–12. Tomsk. (see Krylov, P.N.)

Gadach, E. (1961. History of genus *Ephedra* L. Bot. zh. 49 (2): 243–244.

Garrone, U. 1963. La ripartizione fisico-geographica della Republica Popolare Mongola. Universo, 43 (5): 1010–1011.

Gerasimov, I.P. *1964. Paleogeographic paradox of Pamir.—Izv. AN SSSR, ser. geogr. 3: 4–12.

Goloskokov, V.P. and Kubanskaya, Z.V. 1964. Sympegma formation in Tien Shan. Tr. Inst. bot. AN KazSSR, 18: 3–30. Map.

Golovkova, A.G. *1961. Flora of Central Tien Shan. Mater. k X nauchn. konf. prof.-prepod. sost. Kirgizsk. univ. Sekts. biol. nauk. Frunze: 27–31.

Grubov, V.I. *1964. Flora and vegetation [China]. Central Asian part of country. In: Physical Geography of China. Moscow: pp. 381–438.

———— *1970. Critical notes on taxonomy and nomenclature of some species of genus *Iris* L. in USSR flora. Novosti sist. vyssh. rast. 1969: 29–37.

Grubov, V.I. and Lavrenko, E.M. *1968. In memory of Aleksandr Afanas'evich Yunatov (Dec. 25, 1909–Oct. 24, 1967). Bot. zh. 53 (10): 1493–1500.

Haase, G., Richter, H. and Bartel, H. 1964. Problems of topographic-ecological zonation in the example of Hangay mountains (MNR). Földr. ért. 13 (2): 157–177 (in Hung.).

Hadač, E. 1967. On the history of genus *Ephedra*. Preslia, 39: 5–9 (in Czech.).

Hanelt, P. and Davazamč, S. 1965. Beitrag zur Kenntnis der Flora der Mongolischen Volksrepublik, insbesondere des Gobi-Altai-, des Transaltai- und Alašan-Gobi-Bezirks. Ergebnisse der Mongolischen-Deutschen Biologischen Expedition 1962. Nr. 3. Feddes Repert. 70 (1–3): 7–68. Map.

Hao Hu-tun [Gao Gu-tun'] *1961. Study of eastern boundaries of Qinghai-Tibet upland and arid North-western China in natural zonation of China. Acta Geogr. Sinica, 27: 80–84 (in Chin.).

Hao Shan-u [Gao Shan-u] *1963. Study of sandy desert and Gobi in Sinkiang. Lin'e kesyue Sci. silvae, 8 (1): 56–68 (in Chin.; Russian summary).

Hartmann, H. 1968. Über die Vegetation des Karakorum, I. Vegetatio, 15 (5–6): 297–387.

Hou Hsioh-yu 1961. The concept and fundamental principles of vegetation regions. Acta Bot. Sinica, 9 (3–4): 273–286 (in Chin.).

Hu Chen-hai, Li Huan-min and Li Chen-li [Khu Chzhen-khai, Li Guan-min and Li Chzhen-li].

———— *1964. Distribution and general morphology of Kingdonia uniflora (Ranunculaceae). Acta bot. sinica, 12 (4): 351–358 (in Chin.; English summary).

Hunt, P.F. and Summerhayes, V.S. 1965. *Dactylorhiza nevski*, the correct generic name of the dactylorchids. Watsonia, 6 (2): 128–133.

I Tsao-tsyn. 1962. Some preliminary notes on the complex natural zonation of China. Acta geogr. sinica, 28 (2): 162–168. Schematic map (in Chin.).

Ikonnikov, S.S. *1961, Composition and characteristics of Pamir Flora. Izv. An TadzhSSR, otd. s.-kh. i biol. nauk, 1 (4): 43–53.

———— *1962. Botanical zonation of Pamir. Izv. AN TadzhSSR, otd. biol. 1 (8): 61–45.

———— *1963a. Key to Plants of Pamir. Dushanbe, pp. 1–282. Schematic map.

———— *1963b. About a "near-snow" genus in high-altitude flora. Bot. zh. 48 (1): 85–87. Map.

———— *1970. Additions to Pamir flora.—Novosti sist. vyssh. rast. 1969: 260–275.

Ivanov, A.F. *1968. Utilisation of Desert and Semidesert Pastures of Inner Mongolia. MSKh SSSR. Moscow, pp. 1–107.

Jacot, A.P. 1927. The genus *Gueldenstaedtia* (Leguminosae). J.N. China Branch Roy. Asiat. Soc. 58: 85–125. Map.

Jansson, C.A. 1962. Some species and varieties of *Betula* ser. Verrucosae Suk. in East Asia and N.W. America. Acta Horti Gotoburg. 25: 103–156.

Jurtzev, B. [Yurtzev, B.A.] *1964a. Conspectus of taxonomy of section Baicalia Bge. of genus *Oxytropis* DC. Novosti sist. vyssh. rast. 1964: 191–218.

———— *1964b. Main pathways of evolution of *Oxytropis* of section Baicalia Bge. Bot. zh. 49 (5): 634–647. Maps.

Kalinina, A.V. *1968. Scientific and pedagogic activity of A.A. Yunatov in Mongolia. Bot. zh. 53 (10): 1501–1503.

Kamelin, R.V. *1965. Generic endemism of Middle Asian flora. Bot. zh. 50 (12): 1702–1710.

———— *1969. Materials on Pamiro-Alay flora. *Potentilla biflora* Willd. Bot. zh. 54 (3): 380–388.

Kao Sh-W. 1963. Zum Studium der Sandwüsten und der Gobi in Hsinkiang. Sci. Silv. Peking, 8: 56–68 (in Chin.; Russian summary).

Keng, I.L. and Keng, P.C. 1964. A key to the families of Spermatophyta in China. Rev. Peking, K'o Hsueh Chu Pan She, 1: 108.

Kirpichnikov, M.E. *1965. Publication date of an important work of A.A. Bunge. Bot. zh. 50 (5): 728–729.

Kitagawa, M. 1963. Notulae fractae ad floram Asiae Orientalis (15). J. Jap. Bot. 38 (4): 105–111.

Korovin, E.P. *1961. Experience of phytogeographic zonation of Middle Asia. Tr. Tashkentsk. gos. univ. (186): 25–29.

———— *1962. New genera and species of Umbelliferae flora of Kazakhstan. Tr. Inst. bot. AN KazSSR, 13: 242–262.

Krylov, P.N. 1927–1964.* Flora of Western Siberia. Guide to Identification of Western Siberian Plants. 1–12. Izd. Tomsk. otd. Russk. bot. obshch. i (c 1937) Tomsk. univ. 1 (Pteridophyta—Hydrocharitaceae), 1924: I–X, I–XXIV, 1–138; 2 (Gramineae), 1928: 139–376, I–IX; 3 (Cyperaceae—Orchidaceae), 1929: 377–718, I–XIII; 4 (Salicaceae—Amaranthaceae), 1930: 719–979, I–XII; 5 (Aizoaceae—Berberidaceae), 1931: 981-1227, I–XII; 6 (Papaveraceae—Saxifragaceae), 1931: 1229–1448, I–XII; 7 (Rosaceae—Papilionaceae), 1933: 1449–1817, I–XIV; 8 (Geraniaceae—Cornaceae), 1935: 1819–2087, I–XIV; 9 (Pyrolaceae—Labiatae), 1937: 2089–2401, I–XV; 10 (Solanaceae—Dipsacaceae), 1939: 2401–2627, I–X; 11 (Campanulaceae—Compositae), 1949: 2629–3070, I–XXIV; 12, 1 (Supplement, compiled by L.P. Sergievskaja. Polypodiaceae—Chenopodiaceae), 1960: I–VI, 3071–3255; 12, 2 (Chenopodiaceae—Compositae. Supplement), 1964: 3255–3550, I-L.

Kučera, M. 1968. *Rubus humulifolius* C.A. Mey. in Mongolia. Preslia, 40: 217–218 (in Czech.).

Kuminova, A.V. *1960. Ecogeographic analysis of Altay flora. Vopr. bot. 3: 58–62.

Kurkov, A.A. *1967. Aspects of desert zone formation in the temperate belt of the Northern Hemisphere in the Cenozoic. Izv. Vsesoyuzn. geogr. obshch. 99 (2): 139–141.

Kurochkina, L.Ya. *1962. *Calligonum* in Cherny Irtysh sands (Characteristics of scrub Kazakhstan desert). Tr. Inst. bot. AN KazSSR, 13: 101–132.

———— *1963. Some fragments of vegetation of Central Asian desert in Kazakhstan. Tr. Inst. bot. AN KazSSR, 15: 3–43.

Kuznetsov, B.A. *1963. Data on mammalian fauna of Central Asia. Tr. Mosk. obshch. ispyt. prir. otd. biol. 10: 116–156.

Ladygina, G.M. and Litwinowa [Litvinova], N.P. *1966. Study of plant associations in montane Pamir. Bot. zh. 51 (6): 792–800.

Lauener, L.A. 1963. *Aconitum* of the Himalayas. Notes Roy. Bot. Garden Edinburgh, 25 (1): 1–30.

Lavrenko, E.M. *1965a. Some basic problems in the study of geography and history of plant cover of subarid and arid regions of the USSR and adjacent countries. Bot. zh. 50 (9): 1260–1266.

———— *1965b. Division of Central Asian and Iran-Turan subregions of Afro-Asian desert region into provinces. Bot. zh. 50 (1): 3–15. Map.

———— *1966. Phytogeographic observations in deserts of Kansu corridor and northern rim of Nanshan hills. Bot. zh. 51 (12): 1816–1823.

———— *1968. A.A. Yunatov's contribution to understanding the plant cover of Central Asia. Bot. zh. 53 (10): 1349–1366.

Lavrenko, E.M. and Nikol'skaya, N.I. *1963. Distribution ranges of some Central Asian and North Turan species of desert plants and question of phytogeographic boundary between Middle and Central Asia. Bot. zh. 48 (12): 1741–1761. Maps.

———— *1965. Distribution of some western species of *Stipa* in Mongolian Altay, Junggar and Eastern Tien Shan. Bot. zh. 50 (10): 1419–1428. Maps.

Lavrenko, E.M. and Shul'zhenko, I.F. 1962. Brief review of the Mongolian Agricultural Expedition of the Academy of Sciences USSR, 1947–1952. Izv. Vsesoyuzn. geogr. obshch. 94 (2): 168–175.

Le Tiyan'-yui and Syu Vei-in *1967. Flora of Shensi, Kansu and Ningxia province basins. Izd. kit. inst. lesnogo khoz. Peking: 1–274, ill. (in Chin.).

Li, H.L. 1962. The Chinese species of *Linaria* (Scrophulariaceae). Ac. Sinica Inst. Bot. B. New ser. 3 (2): 205–208.

Li, P.Y. 1965. Additional notes on the *Berberis* species of provinces Shensi, Kansu and northeastern Chinghai. Acta Phytotax. Sinica, 10 (2): 210–214 (in Chin.; Latin diagn.).

Li Shin-in *1961. Basic characteristics of desert plant cover of northern Sinkiang. Acta bot. sinica, 9 (3–4): 287–315. Map (in Chin.; Russian summary).

Ling Jong. 1965a. Genera nova vel minus cognita familiae Compositarum. I. *Vladimiria* Ilj., *Diplazoptilon* Ling et Dolomiaea DC.—Acta Phytotax. Sinica, 10 (1): 75–90 (in Chin.; English summary).

———— 1956c. Genera nova vel minus cognita familiae Compositarum, III. *Syncalathium* Lipsch. Acta Phytotax. Sinica, 10 (3): 283–289 (in Chin.; Latin. diagn.; French summary).

———— 1965d. Notulae de nonnullis generibus tribus Inulearum familiae Compositarum florae sinicae. Acta Phytotax. Sinica, 10 (2): 167–181.

Ling Jong and Chen Li-ling. 1965b. Genera nova vel minus cognita familiae Compositarum. II. *Cavea* W.W. Smith et Small et Nannoglottis Maxim. Acta Phytotax. Sinica, 10 (1): 91–102 (in Chin.; English summary).

Lipschitz, S. [Lipshits, S.Yu.] *1964. Data for genus *Saussurea*, 1. Novosti sist. vyssh. rast. 1964: 314–328.

———— *1966a. Data for genus *Saussurea* L., 2. Bot. zh. 51 (10): 1494–1499.

———— *1966b. Review of species of subgenus *Eriocoryne* (DC.) Hook. f. of genus *Saussurea* DC. Novosti sist. vyssh. rast. 1966: 203–229.

—— *1967. Review of subgenus *Amphilaena* (Stschegl.) Lipsch. of genus *Saussurea* DC. Bot. zh. 52 (5): 651–664.

—— *1968. Critical review of species of section Taraxacifoliae Lipsch. of genus *Saussurea* DC. Novosti sist. vyssh. rast. 1968: 194–229.

Lorch, J. 1962. A revision of *Crypsis* Ait. s.l. (Gramineae). Bull. Res. counc. Israel, sect. D, Bot. 11 (2): 91–110.

Ludlow, F. 1963. A new species of *Corydalis* Sect. Oocapnos (Fumariaceae) from Tibet. Bot. Notiser. 121: 278–280.

Lukanenkova, V.K. *1963. Peculiaritics of Plant cover characteristics at the contact zone of natural regions of Badakhshan, Pamir and Alay (Belyandklik river basin). Bot. zh. 48 (4): 516–525.

—— *1964. South-eastern Pamir as botanical refuge. Bot. zh. 49 (1): 21–29.

Lyu In.-khan. 1962. Some notes on "Problems of Natural Zonation of China". Acta geogr. sinica, 28 (2): 169–174 (in Chin.).

Makoto, T. 1963. Notes on genus *Viola* in Manchuria and Inner Mongolia, 4. *V. xanthopetala* and *V. conferta* in Manchuria and Inner Mongolia. Saisyu to Sinku. Collect. and Breed. 25 (6): 34–38 (in Japanese).

Manibazar, N. 1967. Botanical characteristics and geographic distribution of Mongolian *Adonis* (sp. nova) in Mongolia. BNMAU Shinzhlekh Ukhaany Akad. medee, 4: 42–49. Map (in Mong.; Russian summary).

Marinov, N.A. *1966. Genesis of Eastern Mongolian plains. Izv. Zabaik. fil. Geogr. obshch. SSSR, 2 (3): 129–139.

Merzbacher, G. 1904. Vorläufiger Bericht über eine in den Jahren 1902 und 1903 ausgeführte Forschungsreise in den zentralen Tian-Schan. Peterm. Mitteil. 149: 1–100. Map.

—— 1916. Die Gebirgsgruppe Bogdo-Ola im östlichen Tian-Schan. Abh. Bayer. Ak. Wissensch. math.-phys. Kl. 27 (5): I–VII, 1–330. Maps.

Mesicek, J. and Sojak, J. 1969. Chromosome counts of some Mongolian plants. Folia geobot. Phytotaxon. Tchecosl. 4 (1): 55–86.

Miroshnichenko, Yu.M. *1963. Growing places of some plants in Mongolian People's Republic. Bot. zh. 48 (2): 263–264.

—— *1965. Distribution of *Artemisia frigida* Willd. in Mongolian People's Republic. Bot. zh 50 (3): 420–425.

Misra, K.P. 1963. Phytogeography of genus *Stipa* L. Trop. Ecol. 4: 1–20. Map.

Moore, R.J. 1968. Chromosome numbers and phylogeny in *Caragana* (Leguminosae). Canad. J. Bot. 46 (12): 1513–1522.

Mukerjee, S.K. 1940. A revision of the Labiatae of the Indian Empire. Rec. Bot. Surv. India, 14 (1): 1–228.

Munz, Th.A. 1967. A synopsis of the Asian species of *Delphinium*, sensu stricto. J. Arn. Arb. 48 (3): 249–302, (4): 476–545.

Murzaev, E.M. *1963. Central Asia in the Cenozoic. In: Concepts of Acad. V.A. Obruchev about the Geological Structure of Nor. and Cent. Asia and Their Further Development. Leningrad: 42–61.

—— *1966. Natural Environment in Sinkiang and Desert Formation in Central Asia. Moscow, pp. 1–382. Maps.

Musaev, I.F. *1963. Northern limits of distribution of characteristic components of Turan desert flora. Bot. zh. 48 (2): 157–170.

—— *1966. Geographic distribution of characteristic components of Turan desert flora. Vestn. Leningr. univ. 3: 15–31.

Ochir, Zh. 1965. Vegetation and fodder resources of western part of Hentey upland in MNR. In: Aspects of Meadows and Pastures (Ulan-Bator), 1: 87–112 (in Mong.).

Olziikhutag, N. 1965a. Brief key to families of higher plants of Mongolia. Mongol Ulsyn Ikh Surguul' Erdem shinzhilgeenii bichig [Uch. zap. Mong. gos. univ.] 9 (1): 37–58 (in Mong.; Russian summary).

—— 1965b. Cedar distribution in Mongolian territory. Mongol Ulsyn Ikh Surguul' Erdem shinzhilgeenii bichig, 9 (1): 132–136 (in Mong.).

—— 1966a. New report of two species from Buirnur lake. Mongol Ulsyn Ikh Surguul' Erdem shinzhilgeenii bichig, 10 (3): 37–38 (in Mong.).

—— 1966b. Note on report of *Nymphaea candida* J. et C. Presl. in Mongolia. Mongol Ulsyn Ikh Surguul' Erdem shinzhilgeenii bichig, 10 (3): 54–56 (in Mong.).

—— 1968. Study of the ecology of genus *Oxytropis* in northern Mongolia. Mongol Ulsyn Ikh Surguul' Erdem shinzhilgeenii bichig, 11 (1): 58–62.

Olziikhutag, N. and Tsibukh, V.G. *1968. Distribution of some species of genus *Oxytropis* in Northern Mongolia. Mongol Ulsyn Ikh Surguul' Erdem shinzhilgeenii bichig [Uch. zap. Mong. gos. univ.], 11 (1): 24–35 (in Mong.).

Olziikhutag, N. and Urtnasan, D. 1969. New interesting reports for MNR flora. II. (In: Flora of Hinggan and East. Mongolian provinces). Mongol Ulsyn Ikh Surguul' Erdem shinzhilgeenii SONSGOL, 18: 3–12 (in Mong.; Russian summary).

Olziikhutag, N. and Zhamsran, Ts. 1967. Interesting reports on the flora of Mongolia. Mongol Ulsyn Ikh Surguul' Erdem shinzhilgeenii SONSGOL [Nauchn. soobshch. Mong. gos. univ.], 10: 16–22 (in Mong.; Russian summary).

Pachomova, M.G. (Pakhomova, M.G.) *1967. New species of *Ephedra* from Asia. Bot. mater. Gerb. Inst. bot. AN UzbSSR, 18: 49–54.

—— *1969. Taxonomy of genus *Ephedra* (in connection with works of Yu.D. Soskov and V.A. Nikitin). Bot. zh. 54 (5): 697–705.

Petrov, M. [M.P.] 1962. Types de déserts de l'Asie Centrale. Ann. Geogr. 71 (384): 131–155. Map.

—— *1963a. New geographic compendia on deserts of China. Izv. Vsesoyuzn. geogr. obshch. 95 (4): 380–381.

—— *1963b. Phytogeographic subdivision of deserts of Eurasia and Northern Africa as proposed by E.M. Lavrenko. Bot. zh. 48 (8): 1223–1227.

—— *1963c. Types of Asian deserts. In: Natural Conditions, Animal Husbandry and Fodder base of Deserts. Ashkhabad, pp. 22–48.

—— *1966, 1967. Deserts of Central Asia. 1–2. Leningrad. 1. Ordos, Alashan, Beishan, pp. 1–274, Maps; 2. Khesi, Qaidam, Tarim Basin Corridor, pp. 1–288. Maps.

—— *1969. Topographic zonation of Central Asian desert. Vestn. Leningr. univ. 24: 60–63. Map.

Physical Geography of China. *1964. Moscow, pp. 1–739. Maps.

Pimenov, M. [M.G.] *1965. Species of section Coelopleurum (Ledeb.) M. Pimen., genus *Angelica* L. Novosti sist. vyssh. rast. 1965: 195–206.

Pobedimova, E. [E.G.]* 1966. New species of genus *Hedinia* Ostenf. Novosti sist. vyssh. rast. 1966: 115–121.

—— *1968. A few words on genus *Vincetoxicum* Wolf. Novosti sist. vyssh. rast. 1968: 178–179.

Pojarkova, A.I. (Poyarkova, A.I.) *1950. New fern species and question of Himalayan endemism in the forest relict flora of Middle Asia. Soobshch. Tadzh. fil. AN SSSR, 22: 9–13.

Polunin, O. 1962. Letter from the Karakoram. Quart. Bull. Alpine Garden Soc. 30 (3): 199–213, 228–230.

Potanin, G.N. *1915. Collections of Apollinaria Burdukova on Khankoko mountain range, near Khangel'tsik river, Songin river and from Tsagan-Khairkhan town. Tr. Tomsk. obshch. izuch. Sibiri, 3 (1): 21–22.

Prain, D. 1906. A review of the genera *Meconopsis* and *Cathcartia*. Ann. Bot. 20: 324–370.

Raven, P.H. 1962. The genus *Epilobium* in the Himalayan region. Bull. Brit. Mus. Bot. 2 (12): 325–382.

Raymond, M. 1965. Cyperaceae novae vel criticae. IV. Some Cyperaceae from Karakoram Range (Kashmir). Natur. Canadien, 92 (2): 76–80.

Read, B.E. and Liu, J.C. 1928. Chinese botanical sources of ephedrine and pseudoephedrine. J. Amer. Pharm. Ass. 17(4): 339–344.

Rebristaia, O. [Rebristaya, O.V.] *1964. Genus *Castilleja* Mutis in Eurasia. Novosti sist. vyssh. rast. 1964: 283–311.

Reese, G. 1962. Zur Chromosomenzahl der australischen *Nitraria schoberi* L. Portugal. Acta Biol. Ser. A, 6 (3–4): 295–297.

Roerich, G.N. 1929. Altai—Himalaya. NY.

——— 1931. Trails to Inmost Asia. London.

Roerich [Rerikh], Yu.N. *1960. Acad. N.K. Roerich's expedition to Central Asia (1925–1928). Vopr. geogr. 50: 257–262. Map.

Roldugin, I.I. *1962. Review of Kazakhstan dragonheads. Tr. Inst. bot. AN KazSSR, 13: 263–273.

——— *1968. Pages from the history of restoring the formation of spruce (*Picea schrenkiana* Fisch. et Mey.) in Tien Shan. Bot. mater. Gerb. Inst. bot. AN KazSSR, 5: 20–23.

Sanchir, Ch. 1966a. Stipa of Mongolia. Tr. Inst. biol. AN MNR, 1: 95–104 (in Mong.; Russian summary).

——— 1966b. New species for Mongolia. Tr. Inst. biol. AN MNR, 1: 105–107. (in Mong.; Russian summary).

1967. New flora for Mongolia. Tr. Inst. biol. AN MNR, 2: 191–202. Map (in Mong.; Russian summary).

1968. Floristic characteristics of north-western part of Mongol-Daur hill-forest steppe region. BNMAU Shinzhlekh Ukhaany Akad. medee [Izv. AN MNR], 2: 90–114 (in Mong.; Russian summary).

Sergievskaja [Sergievskaya] L.P. *1966, 1969. Flora of Transbaikal. Izd. Tomsk. univ. 1. Polypodiaceae–Butomaceae), pp. 1–93. Map; 2. Gramineae (grasses), pp. 1–148.

Shilova, N.V. *1965. Wood and leaf structure of *Piptanthus* D. Don and *Ammopiptanthus* Cheng f. Bot. zh. 50 (3): 396–403.

Sidorov, L.F. *1963. Postglacial development of plant cover of Pamir. Bot. zh. 48 (5): 625–638.

Sidorov, L.F. and Potapov, R.L. *1965. History of forests of Pamir and adjoining provinces in the Late Quaternary. Bot. zh. 50 (6): 765–774.

Simizu, T. 1961. Taxonomical notes on genus *Filipendula* Adans. J. Fac. Sci. Techn. Shinshu univ. 26 (ser. A, 10): 1–31.

Simonovicz, L. [Simonovich, L.G.] *1968: Two new species of genus *Adonis* L. from Mongolian People's Republic and China. Novosti sist. vyssh. rast. 1968: 124–129.

Sinitsyn, V.M. *1965, 1966. Palaeoclimate of Eurasia. Parts 1–2. Izd. Leningr. univ. 1. Palaeogene and Neogene, pp. 1–167. Maps; 2. Mesozoic, pp. 1–167. Maps.

Skvortzov, A. [Skvortsov, A.K.] *1966a. Some willows of Indian Himalayas. Novosti sist. vyssh. rast. 1966: 67–74.

——— *1966b. *Salix purpurea* L. and related species. Novosti sist. vyssh. rast. 1966: 48–66.

——— *1968. Willows of the USSR. Taxonomic and Geographic Review. Moscow, pp. 1–262.

Slizik, L. [Slizik, L.N.] *1964. Some critical species of barberry from Middle Asia. Novosti sist. vyssh. rast. 1964: 79–89.

Smirnov, V.A. *1933. Elms of Mongolia. Work of Mongolian Commission. Vestn. AN SSSR, 6: 47–52.

Sojak, J. 1964. *Potentilla tergemina* sp. n. eine neue Art der sibirischen Flora. Preslia. 36: 23–27.

——— 1966. Some new taxa of *Potentilla* L. Folia geobot. et phytotax. bohemoslov. 1 (4): 341–355.

Stepanova, E.F. *1962. Vegetation and Flora of Tarbagatai Mountain Range. Alma Ata, pp. 1–434. Map.

Stevenson, G.A. 1969. An agronomic and taxonomic review of genus *Melilotus* Mill. Canad. J. Plant Sci. 49 (1): 1–20.

Townsend, C.C. 1966. Towards a revision of *Haplophyllum* A. Juss. (Rutaceae), 1. Kew Bull. 20: 89–148.

Tscherneva, O. [Cherneva, O.V.] *1965. New species of genus *Scutellaria* L. from Central Asia. Novosti sist. vyssh. rast. 1965: 220–222.

Tsegmid, Sh. *1962. Physicogeographic zonation of Mongolian People's Republic. Izv. AN SSSR, ser. geogr. 5: 34–41. Schematic Map.

Tsendekhu, Ts. [Tsendeekhuu, Ts.] 1966. Bulbs from Ulan Bator vicinity. Mongol Ulsyn Ikh Surguul' Erdem shinzhilgeenii bichig [Uch. zap. Mong. gos. univ.], 10 (3): 39–42 (in Mong.).

Tserenbalzhid, G. 1966. Plants Growing in Grain Fields in River Basins of Orkhon and Selenga and Some of Their Bioecological Features. Ulan Bator: 1–81 (in Mong.).

–––––– 1967. Data on biology of slender false-pistillate plant. Tr. Biol. inst. Akad. nauk MNR, 2: 289–296.

Tsoon Pu-chiu. 1961. A new system for genus *Pedicularis* (continued). Acta Bot. Sinica, 9 (3–4): 252–274.

Tzvelev, N. [Tsvelev, N.N.] *1964. Genus *Colpodium* Trin. Novosti sist. vyssh. rast. 1964: 5–19.

–––––– *1965. Taxonomy of genus *Calamagrostis* Adans. in the USSR. Novosti sist. vyssh. rast. 1965: 5–50.

Tzvelev, N. [Tsvelev, N.N.] and Grif, V.G. *1965. Karyotaxonomic study of genus *Eremopoa* Roshev. (Gramineae). Bot. zh. 50 (10): 1457–1460. Maps.

Ugrinsky, K.A. [Ugrinskii, K.A.] *1911. Critical notes on some species of Kharkov flora, II. Tr. Obshch. ispyt. prir. Kharkovsk. univ. 44: 287–318.

–––––– 1922. Die Gesamtart *Iris flavissima* Pall. Eine monographische Studie. Beih. Feddes Repert. 14: 1–40.

Ulziikhutag, N. (*see* Olziikhutag, N.)

Vasilevskaya, V.K. and Petrov, M.P. *1964. Central Asian endemic *Tetraena mongolica* Maxim. Bot. zh. 49 (10): 1506–1513. Map.

Vassilczenko I. [Vasil'chenko, I.T.] *1950. History of the origin of *Ephedra* L. Bot. zh. 35 (3): 263–273.

–––––– *1965a. Genesis of genus *Oxytropis* DC. Bot. zh. 50 (3): 313–323. Map.

–––––– *1965b. Continuation of discussion on *Ephedra* L. Bot. zh. 50 (6): 867–870.

–––––– *1970. New species of genus *Oxytropis* DC. in Central Asia. Novosti sist. vyssh. rast. 1969: 152–154.

Vrishch, D.L. *1968. Narrow-leaved lily species of the Far East. Bot. zh. 53 (10): 1466–1475. Maps.

Vvedensky, A.I. and Pazij, V.K. [Vvedenskii, A.I. and Pazii, V.K.] *1967. Notes on Middle Asian species of genus *Eremostachys*. Bot. mater. Gerb. Inst. bot. AN UzbSSR, 18: 8–29.

Wang Chi-wu. 1961. The forests of China, with a Survey of Grassland and Desert Vegetation. Cambridge, Mass.: XIV, 313. Map.

Wang He-shen. *1964. Classification of steppe vegetation of Sinkiang. Sin'tszyan nun'e kesyue, 6: 211–216 (in Chin.).

Wang Wen-tsai. *1962. Critical review of genus *Delphinium* Linn. from Ranunculaceae flora of China. I–II. Acta Bot. Sinica, 10 (1): 59–89; 10 (2): 137–165; 10 (3): 264–284 (in Chin.; Latin diagn.; Russian summary).

Wu Cheng-yih, Li Hsi-wen, Hsuan Shwu-jye, Huang Yong-chin 1965. Materiae ad floram Labiatarum sinensium (1), (2). Acta Phytotax. Sinica, 10 (2): 143–166; 10 (3): 215–242 (in Chin.; Latin diagn.).

Yeo, P.F. 1966. A revision of genus *Bergenia* Moench (Saxifragaceae). Kew Bull. 20: 113–148.

Yunatov, A.A. *1963. Geography and ecology of evergreen desert shrub *Ammopiptanthus* (Maxim.) Cheng. I. Bot. zh. 48 (12): 1804–1812. Map.

Yunatov, A.A. and Banzragch, D. *1968. New data for flora of Northern Hangay. Bot. zh. 53 (10): 1367–1370.

Yurganova, K.V. *1912. From Ulala to Otkhan-Khairkhan (travel diary). Tr. Tomsk. obshch. nauch. Sibiri, 2 (2): 1–38.

Zhamsaran, Ts. *1963. New report of *Polygonatum officinale* All. Sb. nauchn. rabot prepod. i stud. Mong. gos. univ., pp. 10–11.

————— 1965. Key to Mongolian currants. Mongol Ulsyn Ikh Surguul' Erdem shinzhilgeznii bichig [Uch. zap. Mong. gos. univ.], 9 (1): 137–140 (in Mong.).

Zhen Mei-oi and Yan Zhen-tsyan' *1961. Problems of natural zonation of China. Acta geogr. sinica, 27: 66–74. Map (in Chin.; Russian summary).

Zimmerman, J.H. 1958. A monograph of *Veratrum*. Ph. Doct. Diss., Univ. Wisc. (USA), 322 pp. Diss. Abstr. 19: 1536–1537 (1959).

Zimmermann, W. 1965. Zur Taxonomie von *Pulsatilla*. IV. Sechs neue Taxa. Feddes Repert. 70 (1–3): 144–148.

Zimmermann, W. and Miehlich-Vogel, G. 1962. Zur Taxonomie der Gattung *Pulsatilla* Miller. III. Die Subsection Patentes. Kulturpflanze, 3: 93–133.

Zohary, M. 1962. Plant Life of Palestine. I–VI, 1–262. 5 maps in text. NY.

INDEX OF LATIN NAMES OF PLANTS

INDEX OF PLANT DRAWINGS

INDEX OF PLANT DISTRIBUTION RANGES

Index of Distribution Maps

Distribution of *Juniperus, Larix* and *Picea* species given only for non-Soviet part of Central Asia.

Milton Keynes UK
Ingram Content Group UK Ltd.
UKHW020323111024
449327UK00041B/3040